Storey's Guide to Raising Turkeys

THIRD EDITION

Storey's Guide to RAISING TURKEYS

Breeds • Care • Marketing

Don Schrider

Storey Publishing

The mission of Storey Publishing is to serve our customers by publishing practical information that encourages personal independence in harmony with the environment.

Edited by Deborah Burns
Art direction and book design by Cynthia N. McFarland
Cover design by Cynthia N. McFarland, based on a design by Kent Lew
Text production by Theresa Wiscovitch

Cover photograph of Royal Palm turkeys by © Mark J. Barrett/Alamy
Don Schrider's photograph courtesy of the author
Illustrations by © Elayne Sears, except for chart, page 86, by Ilona Sherratt

Indexed by Samantha Miller

Storey Publishing
210 MASS MoCA Way
North Adams, MA 01247
www.storey.com

Printed in the United States by Versa Press
10 9 8 7 6 5

LIBRARY OF CONGRESS CATALOGING-IN-PUBLICATION DATA

Schrider, Don.
Storey's guide to raising turkeys / by Don Schrider. — 3rd edition.
pages cm
Rev. ed of: Storey's guide to raising turkeys / Leonard S. Mercia. c2001.
Includes index.
ISBN 978-1-61212-149-9 (pbk. : alk. paper)
ISBN 978-1-61212-150-5 (hardcover : alk. paper)
ISBN 978-1-60342-909-2 (e-book)
1. Turkeys. I. Title. II. Title: Guide to raising turkeys.
SF507.M47 2013
636.5'92—dc23

2012032591

Contents

Preface

> *"There is really more hope in turkey production for the beginner who starts out by informing himself thoroughly and uses sound judgment in developing his turkey project than for the turkey raiser with years of experience and indifferent success."*
>
> — *From* MARSDEN AND MARTIN'S TURKEY MANAGEMENT, *1945*

THESE WISE WORDS HAVE INSPIRED me throughout my time with poultry. It reminds me that there is advantage in having some flexibility in our thinking and keeping an open attitude to new information. You can say that much of our own success comes from the informed help of others.

My experiences in poultry surely validate this. Lessons from master breeders of poultry have been some of the greatest moments of my life. These men and woman have been generous in sharing ideas with a "young upstart," often with nothing for them to gain. I believe it is for the love of the birds that we find commonality, and it is in the sharing that we fulfill a higher purpose.

I'd like to thank some of my mentors, without whom I would have had very little to share: Richard Holmes, Frank Reese, Jr., James P. Rines Jr., Clint Grimes, Francis LeAnna, Al Watts, Wells LaFon, Donald Boger, Donald Grimes, and John Arbaugh. Additionally, others who may not realize their influence include: Craig Russell, Paul Gilroy, Mark Atwood, Ken Bowles, Danny Williamson, Marjie Bender, Paul Kroll, Miki Adams, Doug Empie, Steve Moise, Alex Hitt, William Yockey, Gay and Rudy Biordi, Tom Walker, Dave Olson, and John Hass.

A host of people made this book possible. Many thanks to my editor, Deb Burns; the gang at Storey Publishing; my editing team — Harvey Ussery, Frank Reese, Paul Gilroy, Michael Focazio, Jim Adkins; my fellow Home Depot employees and managers (who provided me encouragement and time off when I badly needed it), including Bhalla Gaurav, Brandon Hardy, Sandy Rodas, and many others too numerous to list. Thank you to Howard Kogan for pointing me in very good directions. A special thanks to Brenda Powell for doing all those things I could not do while writing and being patient with me. And we cannot forget Jeannette Beranger, whose networking skills are the reason Storey asked me to write this book — please blame Jeannette for any mistakes.

Become an "informed beginner," as Marsden and Martin recommended, and you will surely be successful. I wish you great success with your turkeys.

Lastly, I'd like to dedicate this book to my father, Philip Schrider, my grandmother, Flora Richard, and my poultry mentor, Richard Holmes. There is more of each of you in my writing than you know.

God bless.

— Don Thomas Schrider

If you know turkeys, it will usually follow that you are a successful turkey grower.

— Frank R. Reese, Jr., *master turkey breeder,*
Lindsborg, Kansas

Introduction

MY GRANDMOTHER KEPT A FEW TURKEYS while growing up during the 1930s. They provided a feast for Thanksgiving and Christmas each year, plus income when she sold any extras. "Nanny" Richard kept mostly Bronze turkeys, though she tried Bourbon Reds and Blacks as well. The turkeys ranged around the farm, and her family provided only a minimal amount of grain and dairy products to supplement their diet. One of her fond memories was that the profits from her turkeys earned her a very nice winter coat one year.

Small farms were once commonplace in America. Modern agricultural practices have changed much over the past 70 years — now turkeys are produced by the millions on specialized farms.

A renaissance is underway, however — a renewed interest in keeping poultry in backyards, homesteads, and small diversified farms. Fueling this movement are concerns about food safety and a craving for better flavor and nutrition. For the "king of the poultry yard," the humble turkey, this surge of interest could not have come at a better time.

Turkeys are wonderful and useful creatures. They produce succulent meat, rich eggs for cooking and baking, and nutrient-dense manure. They also can do specialized labor around the farm such as reducing insect populations and fertilizing fields by distributing their manure as they range. Turkeys are curious and social by nature. Imagine a large bird that greets you as you approach and will "answer" you — well, "gobble" at you — when you talk. Their friendly attitude and social nature make turkeys very enjoyable pets.

As people seek appropriate knowledge to guide them in pasture-based production methods, an entire field of "abandoned" knowledge from the first half of the twentieth century is becoming useful again. This book hopes to marry current understandings with time-tested knowledge to give solid information that can benefit turkey keepers of any scale.

The proper way to hold a turkey is to wrap one arm around the bird so that its head faces forward and your hand is supporting the breast. With the other hand, you reach across and hold both of its shanks firmly but gently. In this way the turkey is under control, well supported, and unable to flap its wings, which could cause bruising.

> **READING BODY LANGUAGE**
>
> Standing and stooping are behaviors that indicate a turkey's mood. A sick turkey will tend to stand around rather than sit. It will be lethargic and may have a drooping tail, wings, and head. A turkey unwilling to walk is usually suffering from some ailment and should be checked.
>
> When a turkey is agitated, it will stamp one foot rapidly for just a few seconds. If this agitation has a living source, such as a snake, other turkeys will come to investigate and may take up this stamping too.

The Nature of the Bird

Understanding the nature of the turkey will guarantee success in managing it effectively, regardless of the scale of your enterprise.

Curiosity

Turkeys are extremely curious animals that seem to want to understand every aspect of their surroundings. They appear to have a perfect memory for these details; once they have a working knowledge of something, they seem to categorize it as threatening, of great interest, or of little consequence. This curiosity is more noticeable in heritage and wild turkeys and less so in industrial strains, but it is present to some degree in all turkeys.

Social Network

Being social is another characteristic of turkeys. By nature, turkeys prefer to group together in flocks — also known as **rafters** or **gangs** of turkeys. Each rafter contains a dominant male, lesser males, hens, and poults. The males have a social structure that changes periodically. The female social structure changes very infrequently, but it is important to each turkey to understand its rank and that of the other turkeys. Introducing new turkeys disturbs the social structure; great interest in the newcomer may lead to attempts to drive it away, and even death. Sounds much like a gang, doesn't it?

One male typically fills the dominant role within a rafter of turkeys and enjoys free access to hens. Fights usually occur as the other toms of the flock strive to establish and change the pecking order. The toms will give a stare to anything they intend to attack or to establish or reinforce dominance. Other turkeys prevent conflicts when they evade these "evil eye" stares by avoiding direct frontal alignment, turning away, surrendering their space, lowering their heads, or sitting submissively with neck outstretched pointed away from the aggressor.

Hens will use this stare as well to warn other lower-hierarchy turkeys to yield. Eye contact is an important aspect of turkey nature and comes into play in less-intense interactions as well — in fact, when a turkey sees a turkey it does not recognize it scrutinizes every aspect of the new bird. And it is scrutinized in return.

A turkey will always seek the company of another turkey, and ultimately that of the rafter. When raised by hens, poults tend to stay largely together, almost as if they have an invisible tether preventing them from straying too far. When several poults are lost, each will "lost-call" and then join up with the nearest poults; together they will "lost-call" to find

SECRETS OF A TURKEY GROWER

Bill Barrett was a turkey breeder from Missouri. In a 1951 interview, he made these valuable comments:

- "A turkey is a good deal like a sheep; when something hits him, he seems to give up. He doesn't have any resistance, any fight."
- "The main thing a successful turkey raiser needs is good poults. Then, sanitation and more sanitation and still more sanitation. By sanitation, I mean clean feeders, clean waterers, and clean pens. And, of course, good feed is essential."
- "If there is one critical stage in a turkey's life, it is when he's about 10 to 11 weeks old. That's because turkey men usually shift their birds from the comparative safety of pens to open range at about that age. And it isn't disease so much that bothers them, it's more likely to be weather."

the group. This instinct to stay within a group is a survival instinct deeply embedded in turkeys. It is for this reason turkeys do best in a flock and not alone — though a turkey may become attached to another animal in an attempt to fill this need for companionship. A friend had a tom that decided the chicken hens were his to protect, even from the rooster!

Response to Danger

Turkeys have great eyesight and usually freeze when they see a hawk overhead. They have three levels of reaction to aerial predator threats:

- **First**, a lazy nasal whine that brings attention to the threat.
- **Next**, an ascending purr that causes the flock to freeze or squat. Usually they will then assemble to face the threat.
- **Last**, the turkeys will dive for cover.

Their vision excels in detecting movement. They are especially aware of colors and may even react aggressively to reds and plaids, just as chickens will, interpreting those tones as words.

Getting Around

Walking and trotting are words we have long associated with turkeys — both wild and domestic. Wild turkeys cover a large area each day, walking for miles within a territory. The nonindustrial domestic turkey varieties are very ambulatory as well, often daily exploring the whole of the range allowed to them.

Farmers used to drive turkeys to market, one person leading the way and another at the rear of the flock waving a stick in each hand to encourage the birds in the rear to keep up. Turkeys are very willing to walk and this is an excellent way to move them to the next field.

Turkeys utilize their wedge-shaped bodies when traveling. On range, when a turkey encounters an obstacle such as dense weeds or brush, it uses its body shape to force its way through; its head leads the way, finding an opening, and then its body expands that opening as its legs drive it through. This instinctual behavior can cause problems for the turkey, such as when it encounters a fence — the head makes it through, but the rest of the bird cannot.

Turkeys also take advantage of their shape to ward off rain. They tilt their bodies slightly upward, raise their heads toward the sky, and drop their tails down — presenting as small a surface as they can so the rain rolls off them. This behavior has led to the myth that during a rainstorm turkeys will raise their heads skyward, open their mouths, and drown.

Hygiene Habits

Turkeys are ground-dwelling birds with some healthy adaptations. When just a few days old, they begin dust bathing. This is a social practice, several birds often sharing a choice spot. Dust bathing helps prevent parasite infestations such as mites and lice, removes dry skin, and sometimes cools the bird during hot weather. Dust bathing is often the precursor to

TURKEY LIFE CYCLE

Beards usually do not appear on the gobblers until about October or November, and hens seldom exhibit a beard. The gobbler's beard will continue to grow until about his third year, at which point it seems to have reached full size.

It is not unusual for some toms to wait to breed until their second season — hens typically breed from their first season on. Turkeys have a fibrous, fatty area on their breast, just over the crop, called the breast sponge. This acts as a reservoir of energy. When the toms pay too little attention to food and too much attention to mating, this area will become saggy. This area is also a good indicator of the general condition of turkeys.

Body weight of toms and hens fluctuates throughout the year — reducing as production starts in the spring, continuing to fall through summer, and increasing as fall approaches until maximum weight is achieved about mid-November.

Poults are generally fully feathered by eight weeks of age. From about the fifth week on, caruncles — bare, bright red, fleshy protuberances — begin to appear about the head and throats of the young poults. This stage is referred to as "shooting the red." Young poults are not ready for the pasture until they pass this stage.

feather preening—the act of using the beak to take oil from a gland near the base of the tail and spread this over the feathers.

Breeding Behavior

Like many other birds, turkeys are seasonal breeders. Increasing sunlight stimulates the pituitary gland and causes hormone production and the onset of breeding. Toms become more aggressive to one another, and sometimes to you, and will engage in contests to establish and keep breeding rights.

Toms have bumps, or **caruncles**, on their bare heads and necks, and during the breeding season these areas become bright in color — even changing frequently from powder blue to crimson. The **snood**, the fleshy appendage on the head just above the beak, will elongate and descend, and then contract periodically. You can almost read the mood of a tom by the color of his head and neck and the condition of his snood.

Hens, meanwhile, start to seek out hidden places in which to lay their eggs. Turkey hens are very private and can be very sensitive to disturbances of their nest.

And all this makes sense. In the wild, the colorful tom would attract more attention, and thus predation. A hen needs to seek out a secluded location that the other turkeys will not draw attention to, so that she may sit on her nest of eggs, vulnerable, for 28 days.

Roosting

Turkeys love to roost. It is easy to see that in the wild this protects them somewhat from nighttime predation. Curiously, they seem to prefer open roosts without a roof, even choosing a roost outside during a snowstorm to one under a covered shed. Perhaps it is the fresh air, free of any ammonia vapors, or a built-in mechanism to ensure robust health.

Natural Habitat

The natural environment of the wild turkey is the savannah and open canopy of wooded areas. The natural environment of the domestic turkey is the small farm, with its range and wooded and brushy areas (such as fencerows). Turkeys prefer to have some cover at hand to avoid aerial predators and from which to retreat from the heat of summer days.

◆ ◆ ◆

As you try to understand why turkeys behave the way they do, remember that domestic turkeys still retain much of the nature of their wild ancestors. We can summarize by saying:

- Turkeys are curious, social, and visually stimulated creatures.
- They are sleek and have clean habits.
- They are very ambulatory.
- They are seasonal breeders.
- They like to roost.

If you provide conditions that allow them to express their full nature, they will be more content. If you design housing, equipment, and range that work with their nature, you will be content.

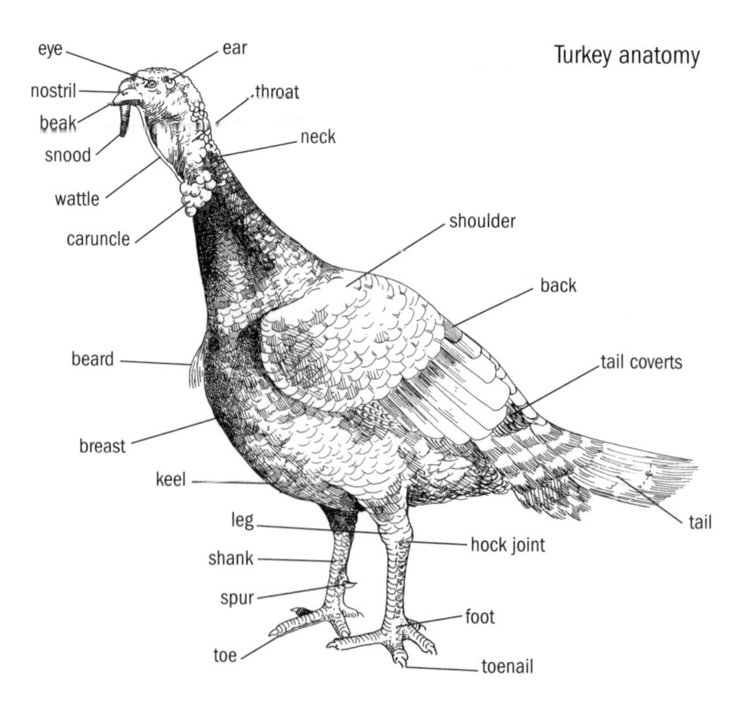

Turkey anatomy

1

History

THE TURKEY IS AN ALL-AMERICAN bird. Estimates place it in the Americas somewhere between 1.8 and 5 million years ago. One of the first animals to be domesticated in this hemisphere, it was also an early partner in nomadic life — providing an excellent food source that was self-sufficient.

An American Original

The Aztecs in Mexico considered the turkey to be an animal of great importance: it was considered one manifestation of the Tezcatlipoca, their trickster god, who held the highest position in the Aztec pantheon. Two religious festivals were dedicated to *Huexolotlin*, the humble turkey, each year, during which eggshells were strewn upon the streets to honor the god that had provided such a plentiful food source. It was not uncommon for a busy Aztec marketplace to see more than 1,000 turkeys sell each day. Meat, eggs, feathers: clearly the turkey played an important role in Aztec civilization.

Other cultures adopted the turkey as well. Even before the Aztecs, the Mayans used turkey feathers, bones, and feet during sacred ceremonies. Today the Navajo consider turkeys a symbol of friendship and plenty, but this was not always the case. They first encountered the birds as raiders of their corn. In an effort to keep the turkeys out, the Navajo fenced the fields. This failed, as turkeys found ways in. The Navajo then changed tactics, opened the pens and fed the turkeys, and domesticated the birds.

NAMED FOR A PEACOCK?

The name *turkey*, attached to these magnificent birds, has three plausible explanations. First, it is believed that Christopher Columbus, thinking he was in India, named them using the Tamil word for peacock, *tuka*. India was, at this period in time, a part of the Turkish Empire; and so *tuka* became turkey. Next is the fact that some North American Indians referred to the bird as *firkee*. Lastly, Luis de Torres, Columbus's physician, is said to have used the Hebrew word for big bird, *tukki*, to describe turkeys. We do know, however, that for more than five hundred years this American icon has been known by the most unlikely of names — turkey.

Many tribes across the Americas hunted turkeys and prized the feathers in the making of special garments. The Wampanoag hunted turkeys with bow and arrow, attracting the birds by imitating their call. In some cases, a child hiding in a pit or behind a log would grab the turkey when it came close. Turkey meat was made into a type of jerky for long preservation. Native Americans even burned woodlands to create meadows to attract and feed turkeys and deer

European Contact

At the time of Cortez, circa 1519, Aztecs were raising turkeys around their homes. It is likely that the flavor of the meat inspired the Spaniards to bring these large birds back to Europe. Cortez and his crew feasted on turkey *mole poblano*, a traditional dish prepared with chocolate, chiles, and turkey. Turkeys first arrived in Spain as early as 1498 on returning vessels. It was not until 1520 that turkeys spread to other parts of Europe. English merchants sampled turkey in Spain and soon imported the birds into England between 1524 and 1541.

The Norse peoples also may have a connection to turkeys. A thirteenth-century painting in St. Peter's Cathedral in Schlesing, Germany, features eight medallions depicting turkeys from "Markland" (America).

Europeans bred turkeys for meat production and the birds became integral members of the farmstead. English farmer Thomas Tusser noted turkey among the farm fare during Christmas of 1573. English navigator William Strickland is credited with introducing the turkey to England in the sixteenth century; a turkey tom is on the crest of his family coat of arms.

Colonial America

Early settlers, from the 1600s on, brought "European" turkeys with them as they colonized the Americas. A 1584 document shows that turkeys were a part of the supplies sent to Jamestown, Virginia, in 1607. It lists "turkies, male and female." Some early varieties brought "back" to America included: Norfolk Black, Black Spanish, Cambridgeshire Bronze, and much later the White Austrian, the Buff, the Blue, and the Ronquieres from Belgium.

The first colonial celebration feast took place at Plymouth Colony in 1621. According to folklore, wild turkey was the centerpiece of the meal and a sign of friendship between Native Americans and the Europeans. The feast became known as an annual "Thanksgiving," appreciating the bounty of the fall harvest — without which people would not survive the winter. By 1700, domesticated turkeys were plentiful in the mid-Atlantic region of the East Coast, and in much of the coastal South.

Breed Improvements

Black and Bronze turkeys were the most common in colonial America. But color played little role until domestic and wild turkeys cross-mated in Rhode Island, producing a color pattern slightly different from the common Bronze turkey. This new variety became known as the Narragansett — named for the nearby bay where they were created. The advantage of having a distinctive color pattern, associated with good meat and flavor, soon made Narragansett turkeys popular.

Rare genetic occurrences, combined with geographic isolation, spawned many wonderful varieties of turkeys. Around 1860, the Rev. R. H. Avery of Wampsville, New York, exhibited a strain of Bronze turkeys that were of a larger size. Bronze turkeys tend to produce occasional offspring with feathers of other patterns or colors, such as white, black, gray, or partridge. By 1874, the American Poultry Association had formed and written standards for several color varieties of turkeys. From this point on, raising turkeys became a profitable farming enterprise.

Breeding for Market

During the early and mid-1900s, turkeys were selectively bred for excellent meat properties, applying old, tried-and-true methods alongside scientific rigor. At that time, turkeys were on pastures all across America, and were naturally mating.

Englishman Jesse Throssel developed Bronze turkeys with broader breasts. Throssel, who moved to Canada in 1926, brought over turkeys in 1927 that he had bred for heavier meat production. Throssel's line is credited with giving rise to the genes needed for broader breasts. Today's broad-breasted commercial turkeys owe their fantastically profuse flesh to his breakthrough in breeding.

Mrs. H. P. Griffin coined the term "Broad Breasted Bronze" to describe these turkeys in 1938. Broad-breasted turkeys became widely popular, dominating the marketplace by the 1950s. During the 1960s, producers began to lose interest in birds with dark pinfeathers, as these sometimes leave a dark, inky blemish on the turkey's skin during processing. Breeders crossed Broad Breasted Bronze turkeys with white turkeys to marry the profuse fleshing with the clean-plucking tendency of white birds. From this point on, white turkeys were the commercial turkey of choice.

During this period, the drawback of breeding turkeys with extreme amounts of flesh became truly evident — short legs, short keel, and profuse flesh affect fertility by preventing natural mating. By the late 1960s, artificial insemination became common practice on commercial turkey breeding farms.

Taking Stock

In 1997, the American Livestock Breeds Conservancy (ALBC) conducted a census of Standard/heritage turkeys. The group found only 1,335 breeding birds total, all varieties included — all that was left of the naturally mating, domestic turkeys. ALBC brought attention to this critically small population, and with partners such as heritage turkey breeder Frank Reese Jr. and Slow Food USA, showed that these turkeys had superior flavor to their industrial counterparts.

In a 2004–2006 partnership with Virginia Tech, ALBC conducted research that showed heritage turkeys have more robust immune systems than their industrial counterparts do. A 2006 census revealed that heritage turkeys are making a comeback, with a breeding population of 10,404 birds.

◆ ◆ ◆

Today turkey is highly regarded for its sumptuous meat — as the centerpiece of celebrations and holiday feasts, and even for everyday fare in turkey sandwiches. The commercial industry produced 248 million turkeys in 2011 — 46 million for Thanksgiving alone. More than 20,000 heritage turkeys were produced in that same year to meet demand.

Turkeys represent a growing market. Now is an excellent time to become a turkey farmer and play a part in the history of this fascinating American bird.

OUR NATIONAL BIRD

On January 26, 1784, Benjamin Franklin wrote a letter to his daughter criticizing the bald eagle and suggesting the turkey would have made a better choice as our national bird.

> For my own part I wish the Bald Eagle had not been chosen the Representative of our Country. He is a Bird of bad moral character. He does not get his Living honestly . . . too lazy to fish for himself, he watches the Labour of the Fishing Hawk; and when that diligent Bird has at length taken a Fish . . . the Bald Eagle pursues him and takes it from him.
>
> I am on this account not displeased that the Figure is not known as a Bald Eagle, but looks more like a Turkey. For in Truth the Turkey is in Comparison a much more respectable Bird, and withal a true original Native of America . . . He is besides, though a little vain & silly, a Bird of Courage, and would not hesitate to attack a Grenadier of the British Guards who should presume to invade his Farm Yard with a red Coat on.

2

Varieties of Turkeys

WHEN WE SPEAK OF TURKEYS, we will find some surprising diversity. Turkeys come in a range of colors and patterns, and a few different shapes as well. As we try to sort through this mix, it is well to first sort things out according to body shape and purpose.

Three Turkey Types

There are essentially three types of turkey:

- The wild turkey, which nature has designed to be independent and self-sustaining
- The heritage or Standard turkey, which farmers have domesticated and bred for meat production and for beauty
- The commercial broad-breasted turkey, which industry has bred to have profuse amounts of flesh, at the cost of natural mating and some mobility

Each of these types has a slightly different structure — and each structure gives those turkeys some advantages.

Selecting the Right Birds

As you choose the right turkeys for your farm or enterprise, you must know the end product you wish to produce, the method of husbandry you expect to use, and your own expectations to match them with the correct type and variety of turkey. For example:

If your market is a high-end restaurant looking to serve game, you should raise wild turkeys (but understand you may need a permit to do so).

If you wish to garner higher prices per pound for heritage turkeys raised on pasture, you should choose a variety of heritage turkey.

If your market wants a truly large turkey, and if you plan to use indoor production, you should consider the commercial broad-breasted variety.

TALKING TURKEY

We usually use the word **variety** instead of breed when we talk turkey. **Breed** is a term usually connected to a turkey's shape and to management of that population. We can say that a breed is a population of domestic animals that share common characteristics and, when bred together, produce offspring with these same characteristics. A breed can reproduce itself.

Wild turkeys are wild and thus are under the management of nature and not man. They are considered a **species** and come in several subspecies.

Commercial turkeys are managed as **industrial strains**, requiring crossing of different strains to unite high reproduction (egg laying) with meat qualities — producing marketable offspring that quickly grow plump. Commercial turkeys are therefore the result of **crossing**, or are **crossbred**.

Neither wild nor commercial turkeys are breeds. That leaves Standard/heritage turkeys, which all share one basic shape or type. The only true breed of turkey is the heritage turkey.

Variety and **subspecies** refer to groups with subtle differences within poultry breeds or species. Wild turkeys come in six different subspecies — with differences such as variations in color pattern, size, and length of leg. Heritage turkeys come in many different varieties — there are eight recognized by the American Poultry Association and more than a dozen others. The differences in varieties usually are connected to feather pattern and color, but may also include flavor or exceptional robustness.

Wild Turkeys

Though few readers are likely to raise wild turkeys, here is a brief description of this type of turkey. Keep in mind that wild turkeys are wild. This means:

- They will not be completely happy unless they have ample room to satisfy their inquisitive natures.
- They are easily startled — a survival trait.
- They can fly.
- They are as likely as not to decide to run away if given a chance.

Wild Turkey (Eastern)

Once called the "forest turkey," this subspecies has the largest range of all wild turkeys, from northern Florida to Maine and all across the eastern half of the United States, up into the Canadian provinces of Manitoba, Ontario, and Quebec. Males can reach 30 pounds (13.6 kg) in size and 4 feet (1.2 m) in height. They have tail coverts tipped with chestnut brown.

Gould's Wild Turkey

This turkey subspecies is native to Arizona, New Mexico, and the central valleys and northern mountains of Mexico. They are the largest of all subspecies, with longer legs, larger feet, and longer tail feathers. Body feathers are copper and greenish gold in iridescence.

Merriam's Wild Turkey

This subspecies ranges across the prairies of Montana, South Dakota, and Wyoming, as well as through the Rocky Mountains and high-mesa country of New Mexico; there's also a pocket in Idaho and Oregon. Merriam's Wild Turkeys prefer mountainous areas with ponderosa pine. The tails and lower back feathers of this subspecies have white tips and purple and bronze iridescence.

Osceola Wild Turkey

Also known as the Florida Wild Turkey, this bird ranges across the Florida peninsula. Its body feathers are green-purple in iridescence, the wing feathers are dark with small amounts of white barring, and it is smaller and darker than the Wild Turkey (Eastern).

Rio Grande Wild Turkey

This subspecies ranges along the Rio Grande River; it is also found in Colorado, Kansas, New Mexico, Oklahoma, Oregon, and Texas. It has been introduced to a few eastern states, to central and western California, and to Hawaii. These turkeys have long legs and are adapted to prairies. Their body feathers have a green coppery sheen, their lower back feathers are buff to very light tan, and the tips of their tail feathers are buff or light tan.

South Mexican Wild Turkey

Considered the nominate — or first-named — subspecies, South Mexican Wild Turkeys range in the southern expanses of Mexico and were the first wild turkeys encountered by Europeans. The Aztecs domesticated this subspecies, thus giving rise to domestic turkeys. The tame descendants of the South Mexican Wild Turkey traveled with Spanish explorers back to Spain, and later spread throughout Europe.

Commercial Broad-Breasted Turkeys

There are a small number of different strains of commercial turkeys. Industry closely guards their genetics, so the primary strains are very difficult to obtain. Poultry hatcheries do, however, make available some of these broad-breasted genetics.

You may order Broad Breasted White or Broad Breasted Bronze turkeys from hatcheries. Expect the poults to grow into very large turkeys that will be ready for processing within 16 to 20 weeks. Feed conversion will

PRIMARY COMMERCIAL STRAINS

Nicholas: owned by Aviagen, available in the United States
Hybrid: owned by Hybrid Turkeys of Canada, available in the United States
British United Turkeys (a.k.a. BUT): owned by Aviagen, available in the United Kingdom and Europe

be excellent. These turkeys cannot mate naturally, so if you plan to breed them, see the section on artificial insemination in this book (page 221).

Heritage Turkey Varieties

These are the old varieties of domestic turkeys that played an important role on farms from the colonization of America up through the early 1950s. They produce more meat than wild turkeys, but grow slower and produce less meat than the commercial broad-breasted turkeys. They strike a nice balance between meat production and natural mating, and are designed for range conditions. Be aware, though: many strains have been greatly neglected or poorly selected for some time, so quality will vary as much from flock to flock as between varieties.

WHAT THE "OLD" EXPERTS KNEW THAT YOU SHOULD KNOW

"I was friends with many people through the years that I considered to 'really know their Standard-bred turkeys.' What did these 'Old Experts' know? I write this from the knowledge learned from past turkey growers who raised only Standard-bred turkeys and not the commercial industrial turkeys. I also write from my personal 50 years of raising Standard-bred turkeys and a few commercial White and Bronze turkeys.

"The rearing of hybrid turkeys and their ability to handle stress factors is much different than with Standard-bred turkeys. The commercial turkey's diminished immune system is very different from that of the old Standard-bred turkey, which still has a healthy immune system. So the stress factors and their effects can be a bigger problem with the commercial poult.

"If you remember nothing else, remember that commercial Broad Breasted White or Bronze turkeys were not engineered for self-breeding or pasture rearing. If you attempt to rear them out on pasture, you are placing a tremendous amount of stress on them that they were not designed for."

— Frank R. Reese, Jr., *master turkey breeder and owner,*
Good Shepherd Turkey Ranch, Tampa, Kansas

The American Poultry Association (APA) has, since 1874, governed the standards by which these turkeys are selected. There are eight varieties of heritage turkeys recognized by the APA: Beltsville Small White, Black, Bourbon Red, Bronze, Narragansett, Royal Palm, Slate, and White Holland.

Other heritage varieties have long histories but have not sought Standard recognition. Many subtle color variations also occur when some turkey varieties are crossed or when a variety is lacking or weak in certain genes. Nevertheless, all heritage turkeys share a basic shape, or type; the differences are largely cosmetic in nature.

The following is a list of descriptions that includes the APA-recognized varieties as well as some of the unrecognized varieties.

Standard Heritage Turkeys

The following are varieties of heritage turkeys that have been recognized by the American Poultry Association, many of which were economically important up until the 1950s.

Beltsville Small White

ORIGIN. Recognized by the American Poultry Association in 1951, the Beltsville Small White is the result of the work of the USDA research center in

Beltsville Small White

Beltsville, Maryland, responding to consumer demand in a 1936 survey for a small, meaty turkey. Between 1934 and 1941, researchers crossed Bronze, Wild Turkey (Eastern), Narragansett, White Austrian, and White Holland turkeys. The bird was in use by the 1940s and by the mid-1950s was extremely popular. But its small size—just right to fit into a small refrigerator—became its downfall as America prospered and families sought larger turkeys. By the 1970s, the Beltsville Small White turkey was nearly extinct.

DESCRIPTION. These turkeys have white plumage, black beards, horn-colored beaks, and dark brown eyes. Standard weights are 21 pounds (9.5 kg) for males and 12 pounds (5.4 kg) for females.

Black

ORIGIN. Recognized by the American Poultry Association in 1874, Blacks developed from the original stock taken to Europe in the 1500s. Black turkeys have been called other names, such as Norfolk Black in England and Black Spanish, to differentiate the sources of lines. In America, however, these lines have been crossed and recrossed for so long that it is only proper to refer to any population as Black turkeys. Many sources also suggest that Wild Turkey (Eastern) has been introduced into the Black turkey lines during their history. Up until the 1960s, this variety had been the fifth-most-popular variety and was a favorite in Maryland and Virginia. It placed very high in tasting competitions and comparisons.

DESCRIPTION. Black turkeys are very beautiful, with green-iridescent black feathers covering their entire body and tails. As poults, they may grow some white or bronze feathers, especially on their heads, but these will molt out as they grow. Standard weights are 33 pounds (15.0 kg) for males and 18 pounds (8.2 kg) for females.

Bourbon Red

ORIGIN. J. F. Barbee developed this variety in Bourbon County, Kentucky, during the late 1800s, when he crossed a Tuscarora—a dark, reddish buff turkey from Pennsylvania — with Bronze and White Holland turkeys. The American Poultry Association recognized Bourbon Red in 1909.

It was the fourth-most-popular variety and was very popular in its native Kentucky and across the United States, up until the 1940s. It has won several tasting competitions and was the first heritage turkey variety to gain the attention of national media around 1998.

DESCRIPTION. Bourbon Reds have a rich, deep red color over most of their feathers, with white tail and white wing feathers. Standard weights are 33 pounds (15.0 kg) for males and 18 pounds (8.2 kg) for females.

Bronze

ORIGIN. Recognized by the American Poultry Association in 1874, the Bronze originated from crosses between domestic turkeys and wild turkeys in colonial America. Occasionally this variety produces offspring of other colors, such as white, black, or partridge, and it has given rise to the other varieties. The Bronze was the most popular variety of turkey up until the 1960s, commonly found all over the United States and Canada. It placed very high in tasting competitions and comparisons.

DESCRIPTION. The name "Bronze" comes from the coppery iridescent sheen of its body feathers, but these turkeys did not acquire the name until the 1830s. Standard weights are 36 pounds (16.3 kg) for males and 20 pounds (9.1 kg) for females.

Narragansett

ORIGIN. Recognized by the American Poultry Association in 1874, this variety developed from crosses of domestic turkeys, brought to America by colonists, with Wild Turkeys (Eastern). It derives its name from the area around Narragansett Bay, Rhode Island, where the variety originated. Narragansett turkeys were very popular for meat quality and ability to raise their own chicks — it was not unusual for a dozen hens to produce and raise more than one hundred poults with just a little supplemental grain. It has been the third-most-popular heritage turkey variety and was very popular across New England and the mid-Atlantic.

DESCRIPTION. Narragansetts have a unique color pattern that is a blending of black, gray, tan, and white, with a steel-gray iridescence to their body feathers. This color combination is not found on turkeys outside North America. Standard weights are 33 pounds (15.0 kg) for males and 18 pounds (8.2 kg) for females.

Royal Palm

ORIGIN. Enoch Carson of Lake Worth, Florida, developed this variety during the 1920s. The American Poultry Association recognized Royal Palms in 1971. They appear to be descendants of crosses of Bronze, Black, and Narragansett turkeys with Osceola Wild Turkeys. Other turkeys of unrefined versions of this coloration have been known to occur since the 1700s, and in Europe are known under different names, such as Black-Laced White, Crollwitz, and Pied. This variety has never been popular for meat production; it is valued instead for its beauty. Royal Palm turkeys are a little lighter in weight than the other heritage turkeys, are very active, and are better suited to the hot, humid climate of Florida and the southern United States — all of these traits likely due to their wild ancestor. They can win tasting competitions for flavor, regardless of their lighter frame.

DESCRIPTION. Royal Palm turkeys are a beautiful, clean white with black bars across their feathers (see a color photograph on the front cover of this book). Standard weights are 22 pounds for males (10.0 kg) and 12 pounds (5.4 kg) for females.

Slate

ORIGIN. Recognized by the American Poultry Association in 1874, Slate turkeys were raised up through the 1940s but they were never as popular as other varieties. In areas where several turkey farms ranged their flocks, Slate turkeys' color made them easily distinguishable. This variety has an extremely robust immune system, according to tests conducted from 2004 to 2006 by Virginia Tech, in partnership with the American Livestock Breeds Conservancy.

Slate

SLATE GENETICS

Slate coloration is the result of two different mutations — one recessive and one dominant — resulting in several subtle variations. This color has been around a very long time, and no one is quite sure when or how it developed. One theory suggests that a foundation of Black turkeys crossed with White Hollands.

DESCRIPTION. Slate turkeys are a bluish slate color with black and sometimes white flecks. This coloration is lighter in females and can range from a very light shade, called self-blue, to a fairly dark shade. It may also be called Blue, Blue Slate, Lavender, and Slate Blue. Lilac and Red Slate turkeys are variants of Slate turkeys with the addition of some red. Standard weights are 33 pounds (15.0 kg) for males and 18 pounds (8.2 kg) for females.

White Holland

ORIGIN. The name "White Holland" implies Dutch origins, but white turkeys have been around for a very long time — the Aztecs selectively bred white turkeys. White Holland turkeys descend from white turkeys

brought to America by colonists who crossed them with the white offspring of Bronze turkeys and any other white turkey that happened along. The variety is also the ancestor of today's commercial turkeys, giving them the white coloration that allows for clean plucking. The American Poultry Association recognized White Hollands in 1874. This turkey was the second-most-popular variety of turkey up through the 1950s, especially in New England.

DESCRIPTION. White Holland turkeys have pure, snow-white feathers, black beards, pink to horn-colored beaks, and brown eyes. Standard weights are 36 pounds (16.3 kg) for males and 20 pounds (9.1 kg) for females.

Non-Standard Heritage Turkeys

The following are varieties of heritage turkeys that are not currently recognized by the American Poultry Association, often because of low population size. Some of these varieties were once quite populous or economically important.

Auburn

ORIGIN. This is a very old turkey variety that lacked popularity but has survived for several hundred years. As far back as the late 1700s, there is record of Auburn turkeys trotting into the Philadelphia market. Auburn turkeys are essentially a Bronze turkey with a recessive gene that causes the black sections to turn brown—just as the dun gene turns black feathers into chocolate. The variety was originally named Brown, but in 1990 the name was changed to Auburn to better describe the plumage, which has a pattern (like that of the Bronze), rather than one even shade.

DESCRIPTION. Auburn turkeys are similar to Bronze turkeys in appearance, except the black and iridescent feathers are a dull brown color. Weights are 25 pounds (11.3 kg) for males and 16 pounds (7.25 kg) for females.

Chocolate

ORIGIN. Turkeys of this coloration were once popular in the southern United States. Chocolate turkeys were also found in the south of France, and so it is believed that the French exported these to the American

Chocolate

South. The Civil War nearly wiped out the population of Chocolates, and only in recent years has the variety seen any kind of resurgence.

DESCRIPTION. As their name implies, Chocolate turkeys have dark chocolate–colored plumage, legs, and feet. They are essentially a Black turkey with a recessive dun gene. A Chocolate tom bred to Black hens will produce pure Chocolate daughters but the sons will be impure for the Chocolate color. Since the Civil War, breeders — with little genetics knowledge to go on — have introduced impurities into the variety with Bronze, Narragansett, and even red genes. Off-colored feathers and offspring of other color patterns, like Auburn turkeys, are indicators of impure ancestry. Chocolate turkey weights are 23 pounds (10.4 kg) for males and 14 pounds (6.3 kg) for females.

Jersey Buff

ORIGIN. The Buff turkey was recognized by the American Poultry Association in 1874, but dropped due to lack of interest. In the 1940s, the New Jersey Agricultural Experiment Station in Millville decided to re-create Buff turkeys. They used Bourbon Red, Black, and Broad Breasted Bronze turkeys (early Broad Breasted Bronze that were still naturally mating). So Buff turkeys created Bourbon Red turkeys, which, in return re-created Buff turkeys. According to the National Poultry Improvement

Jersey Buff

Plan, Jersey Buff turkeys were the fourth-most-popular variety in 1951 to 1952 with more than 25,000 in the population. They were popular in New Jersey, eastern Pennsylvania, and California.

DESCRIPTION. Males and females vary in shade of coloration, males being darker, but both are a rich, even buff color with very light buff or white in wings and tails. Weights are 21 pounds (9.5 kg) for males and 12 pounds (5.4 kg) for females.

Midget White

ORIGIN. Dr. J. Robert Smyth created this variety during the 1960s at the University of Massachusetts by crossing Broad Breasted White and Royal Palm turkeys. Smyth's goal was creation of a smaller version of the broad-breasted commercial turkey to meet an anticipated demand that did not materialize. Soon after creation, the flock was dispersed. Dr. Bernie Wentworth, a former graduate student of Smyth's and a professor at the University of Wisconsin, was surprised to find some of these turkeys with University of Massachusetts bands in a backyard flock. He acquired some for the University of Wisconsin's flock and bred them until his retirement in the late 1990s, when they were again dispersed.

DESCRIPTION. Aptly named, these small birds have white plumage. Midget Whites are heavy egg producers, laying 60 to 80 large eggs per year with high hatchability. They have a fast rate of growth, and, though broad-breasted, mate naturally. Males weigh 13 pounds (5.9 kg) and females weigh 8 pounds (3.6 kg).

Mottled Black

ORIGIN. This variety of turkeys was developed in the mid-1990s by Kevin and Jill Porter of Indiana, who wanted to produce feathers that resemble those of immature Bald Eagles for legal use in arts and crafts.

DESCRIPTION. The turkey is black with many mottles of white on its feathers and some white feathers with specs and spots of white. Mottling increases as the birds age. Weights are 23 pounds (10.4 kg) for males and 14 pounds (6.3 kg) for females.

Regal Red

ORIGIN. Solid red turkeys were often called Arkansas Red or Kentucky Red. Bourbon Red and Buff turkeys have historically produced occasional offspring that are nearly solid red in color. In 2000, Dr. Tom Walker of Texas, remembering solid red turkeys, set out to revive these turkeys of his youth. Using Red turkeys he secured from another breeder, Walker crossed with Black turkeys, in order to improve vigor, and made selection decisions based on a combination of poult down color and adult wing and tail color.

Regal Red

DESCRIPTION. Regal Red poults have pure-white down at hatch but grow into solid red birds by maturity. Weights are 23 pounds (10.4 kg) for males and 14 pounds (6.3 kg) for females.

Sweetgrass

ORIGIN. Perhaps one of the most beautiful of turkey varieties, the Sweetgrass is very difficult to obtain. In 1996, tricolored turkeys appeared from the naturally mating Broad Breasted Bronze flock of Sweetgrass Farm in Big Timber, Montana. Sweetgrass turkeys are named for this occurrence.

DESCRIPTION. The Sweetgrass color pattern seems to arise from a combination of genes that also cause Royal Palm and Black-Winged Bronze color patterns. Sweetgrass turkeys clearly resemble a mix of the Royal Palm and Bronze color patterns with a tinge of red throughout. Turkeys of this color have appeared in the past, under such names as Calico, Tricolor, and Yellow-Shouldered Ronquière. In fact, such birds appear in a 1566 painting by Flemish master Joachim Beuckelaars. Weights vary greatly from strain to strain and are 17 to 25 pounds (7.7–11.3 kg) for males and 12 to 16 pounds (5.4–7.2 kg) for females.

Other Variations

Many other non-Standard color variations occur in turkeys as well. In Bronze you find variants such as Black-Winged Bronze and Crimson Dawn that have subtle differences from the standard Bronze coloration. Silver appears in a few turkey varieties, such as Silver Auburn or Silver Dappled. There are various blue and red versions of established varieties,

AVOID BRINGING HOME PROBLEMS

Avoid auctions and securing live adult and juvenile birds from several locations — there is simply too great an opportunity for exposure to diseases. Swapping birds is also a bad idea. Bringing in one seemingly healthy bird could expose your entire flock to a disease that can lead to health issues for generations — or even mass deaths. If you secure stock from a master breeder, return to only this one source for new stock.

such as Blue Palm, Red Palm, Chocolate Palm, Blue Red Bronze, Red Bronze, and Red Slate. There is even a Penciled Palm — a combination of a Royal Palm color with some penciling throughout the body feathers, wings, and tail. The possible color combinations seem nearly endless, but few of these exist in populations of any size.

Securing Stock

Now that you have seen the great variety of turkeys available out there, it's time to select a variety and secure stock. As I will say repeatedly in this book, try to match the turkey variety to the product you intend to produce with the system of production you intend to use. You should select a variety that is available and that you find appealing — after all, you will need to look at the turkeys every day.

Ask Questions

In finding a source of stock, I recommend a few qualifications to help guide you to a good choice.

First, make sure you're selecting disease-free stock. The best method of ensuring the parent stock is disease free is participation in the U.S. Department of Agriculture's National Poultry Improvement Plan (NPIP). Ask the hatchery or producer for their NPIP number and find out which diseases the breeder tests for and the frequency.

Next, determine if the breeder can provide annually the number of poults you need, and at the time of year you need the stock. Too many people think they can purchase poults "right off the shelf," when in fact, the breeder needs several months' notice — count back from your delivery date to include 4 weeks for hatching, 2 weeks for collecting eggs, and 2 to 4 weeks to feed a breeder diet and to use lights to bring the hens into production. In many cases, stock may not be available because it is already reserved for another customer. Plan ahead and place your order as early as you can.

Next, ask the source how long they have had the variety. This question gives you a good idea of the source's knowledge and the likely quality of the stock. Ask about selection methods. Likely, they will have little information here, but you could stumble upon a master breeder with a surprising wealth of information on the stock — BINGO! You just scored good stock and a possible mentor.

Lastly, ask for the origin of the stock. Many breeders are conserving lines and strains that have been handed down for generations, and it is a blessing to know the history of your birds at the start.

Some May Be Rare

For some varieties, the stock you want will be in the hands of only a small breeder, or a few small breeders, spread across the country. It may be hard to get the number of birds you need, and quality may vary greatly. If you plan to maintain your own flock of breeder birds — and I heartily suggest you do — secure as many poults as you can from each source to start with, cull down to the best from each source, and begin a breeding program. Within 3 years, you will have the quality and consistency you need. Within 5 years, you will have raised the quality of your strain to make the birds worthy and productive.

Shipping and Receiving

Eggs and day-old poults can be shipped through the U.S. postal system, a very affordable way to secure a start in stock. Even if you plan simply to order poults each year, this is still a viable method.

To be sure everything goes smoothly, find out from the hatchery the expected delivery date and let your post office know live poults are on their way. Ask the shipper to put your phone number on the box and ask the postal clerk to call you as soon as the poults arrive — this may save several hours of stress for the poults and allow you to get them settled sooner.

Have your brooder running and up to temperature for a few days prior to the poults' arrival. This way you avoid chilling already stressed poults.

In some cases, you may order adult stock and have it shipped across the country. For shipped stock, ask the breeder to include a few quartered apples when the birds are sent; the fruits are a good source of energy, nutrition, moisture, and relief for boredom. If you do have birds shipped, avoid the warm months of summer, from June through September, so that the turkeys do not risk heat stroke.

REAL SURVIVORS

During the early twentieth century, game managers estimated the total population of wild turkeys to be approximately 30,000 birds. Barely hanging on in remote pockets of the United States, by the 1940s the wild turkey was nearly hunted to extinction in Canada. By limiting hunting, game officials allowed the population to rebound on its own. By 1973, the estimated U.S. population of wild turkeys was at 1.3 million. Today's current estimates places the total population at 7 million.

Heritage turkeys were down to a population of 1,335 breeding birds in 1997. Thanks to an expanding market that valued the robust flavor of their meat, by 2006 the breeding population had climbed to 10,404 birds.

Adult or poult, the first thing to do when shipped birds arrive is provide them with water. After an hour, or so, feel free to feed, but do not feed them until they have had time to drink. This avoids dehydration — the food takes away moisture as it is digested — by ensuring they first can replenish the water they have lost. More shipped poultry die of dehydration than from any other cause. (See pages 105–106 for more information.)

◆ ◆ ◆

There are many different kinds of turkey. Take some time and pick a variety that suits your needs, find a source for the stock, plan for the day of arrival, and begin your journey into raising turkeys!

3

Buildings and Equipment

WELL-DESIGNED BUILDINGS AND EQUIPMENT will help ensure a successful turkey operation. Equipment for turkeys need not be complicated, expensive, or purchased. The most important point with buildings and equipment is that the nature of turkeys has been considered in the design. This chapter provides some equipment and building designs that have proven effective for other turkey growers when utilized properly.

Management Systems

Several types of management systems may be used successfully to grow turkeys. The majority of commercial producers now grow their turkeys in confinement. The poults are usually brooded in the turkey house and stay in the house until marketed. They are frequently started in a small portion of the house and, as the birds become larger and need more space, gradually permitted to use the whole house.

Porch-rearing was once a popular method. In this system, porches are attached to the brooding facility. When the birds reach a few weeks of age, they are allowed to go out on the porch area for a few weeks before being put out on range. Some are raised to market age on the porches. This is still a viable way of growing small flocks of turkeys.

Another method used, where suitable land is available, is range-rearing. The birds are started in a brooding facility and, at roughly six to eight weeks of age, are put out on range for the remainder of the growing period, depending on the weather.

There are also variations of some of these methods that small-flock producers use successfully. Remember that none of these methods is mutually exclusive. You can use any one method as described, or you can borrow from all of them to create your own unique system.

Porch-Rearing

Turkeys that have never been outside the brooder house may not seek shade from sunlight or shelter from the rain when placed on range. For this reason, producers who plan to range their turkeys frequently put young birds on sun porches attached to the brooder house. When the weather is warm, the young turkey poults can leave the brooder house and go out on the porches as early as three weeks of age. Usually, the porches are covered with fine-mesh woven wire on the sides and top to prevent the poults from getting out. The floors may be either slat or wire. Either one works well if the porches will be used for just a few weeks before the birds go on range. However, if the birds are to be raised on porches up to market age, wire floors are not satisfactory, particularly for heavier turkeys. Some birds tend to develop foot and leg problems as well as breast blisters or sores.

Smaller varieties of turkeys and those to be dressed at an early age for broilers or fryers do quite well on wire floors. Grow larger varieties for heavy roasters on porches with slat floors.

If you locate the feed and watering equipment so that you can service them from outside the porch, you will greatly simplify the chores. (For more on sun porches, see page 106.)

CHICKENS VS. TURKEYS

To prevent blackhead disease, keep chickens at least 200 feet (61 m) away from the turkey barns and pastures; use a barrier. Chickens must have their own pasture, barn, watering equipment, and feed equipment. See chapter 12 for more on blackhead.

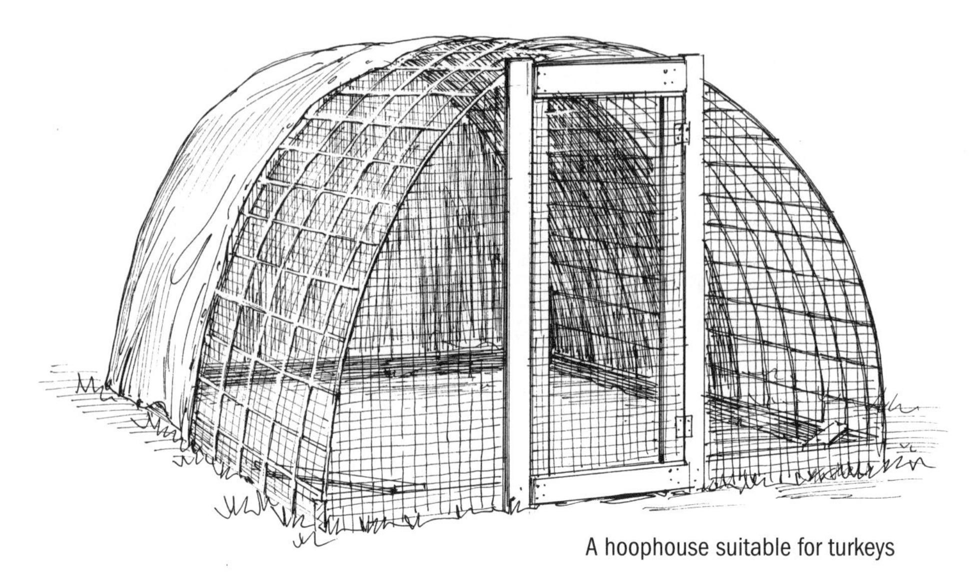

A hoophouse suitable for turkeys

Range-Rearing

Range-rearing offers an opportunity to reduce the cost of growing turkeys. This is especially true if the diet can be supplemented with homegrown grains. Turkeys are good foragers. And if good green feed is available on the range, this means less consumption of expensive mixed feed, thereby reducing the cost of the feeding program. Building costs are much lower when birds are range-reared, but labor requirements are higher. (See chapter 7 for more information.)

Portable Range Shelters

Depending on climatic conditions, some growers provide only roosts for turkeys on range. Others actually allow the turkeys to sleep on the ground. This method is more practical when the turkeys will be matured early and before the cold winter weather sets in. Portable range shelters give the turkeys much better protection during poor weather. They can be moved to new locations to provide the birds with better range conditions and prevent development of muddy spots and contaminated areas.

When portable shelters are used with roosting quarters, the feeders and waterers can be moved whenever the grass is closely grazed in an area. Commercial producers sometimes provide pole buildings for shelter at night and let birds out on range during the day.

CONSIDERATIONS FOR PORTABLE RANGE SHELTERS

Make sure there is enough space for all birds to get into the shelter at one time.

- Typical dimensions of portable shelters are 10 × 12 feet (3 m × 3.7 m) or 12 × 14 feet (3.7 m × 4.3 m), but they can be built smaller to accommodate small flocks. If built any larger, they are not as easily moved, and there is greater chance for building damage during a move.
- A 10 × 12-foot (3 m × 3.7 m) shelter can supply roosting space for up to 60 twelve-week-old turkeys and up to 30 mature birds.

Curtains that roll down from the top might be installed on one or two sides to block rain from the prevailing winds during stormy weather. This is especially important if the weather is bad shortly after young birds are moved onto a range.

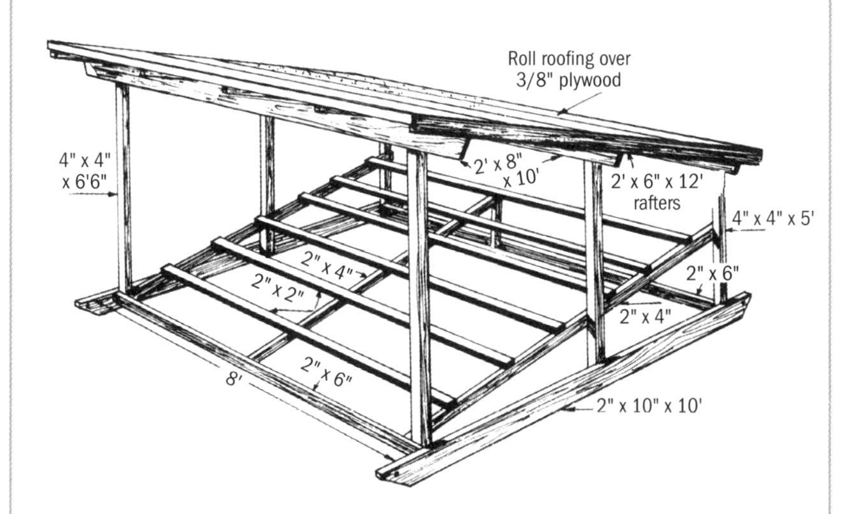

Portable turkey range shelter

Precautions and Special Considerations

Normally, May- and June-hatched poults can be put out on range at eight weeks of age. Before putting them on range, make sure they are well feathered, especially over the hips and back. Check the forecast, and try to move them out during good weather. It is best to move the birds in the morning to give them time to adjust to their new environment before darkness.

If possible, provide a range area that has been free of turkeys for at least 1 year and preferably 2 years. You can use a temporary fence to confine the flock to a small part of the range area. Move the fence once a week or as often as the range and weather conditions indicate.

Provide artificial shade if there is no natural shade. Several rows of corn planted along the sunny side of the range area provide good shade and some feed. If range shelters are used, move them every 7 to 14 days, depending on the weather and on the quality of the range. Move the feed and watering equipment as needed to avoid muddy and bare spots.

Range Feeders and Waterers

Range feeders should be waterproof and windproof so that the feed does not spoil or blow away. Place the feeders on skids or make them small enough that they can be moved by hand or with the help of a small tractor. Trough-type feeders are inexpensive and relatively easy to construct. Specialized turkey-feeding equipment is also available.

Range feeders can be made of wood, but must be easy to fill and able to protect the feed from wind and rain. Move them every day or two to prevent muddy spots.

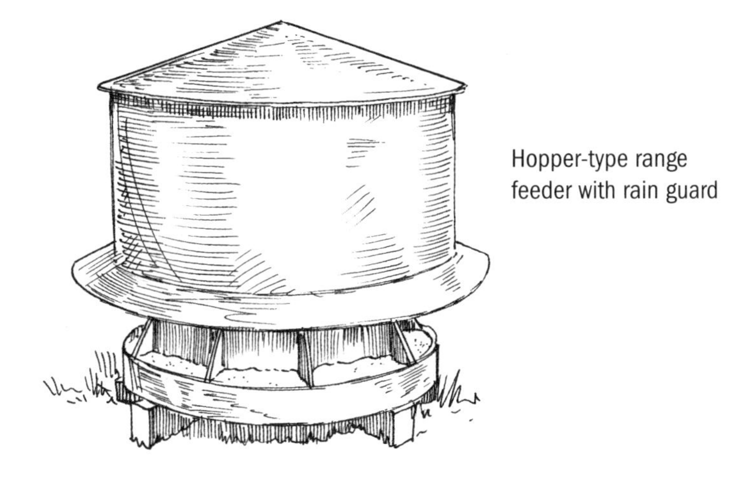

Hopper-type range feeder with rain guard

To minimize waste, all feeding equipment should be designed so that you can adjust it as the birds grow; the lip of the feed hopper should be approximately in line with the bird's back. For the same reason, the feed hopper should never be more than half full. Pelleted feeds are less likely to be wasted on range. Provide at least 6 inches (15.2 cm) of feed trough per bird if the feeders are filled each day. Feeders with storage capacity require less space, and the amount of feeder space should conform to the equipment manufacturer's recommendations.

Clean the waterers daily and disinfect them weekly. Locate waterers close to the shelters. If possible, shade the waterer with portable or natural shade.

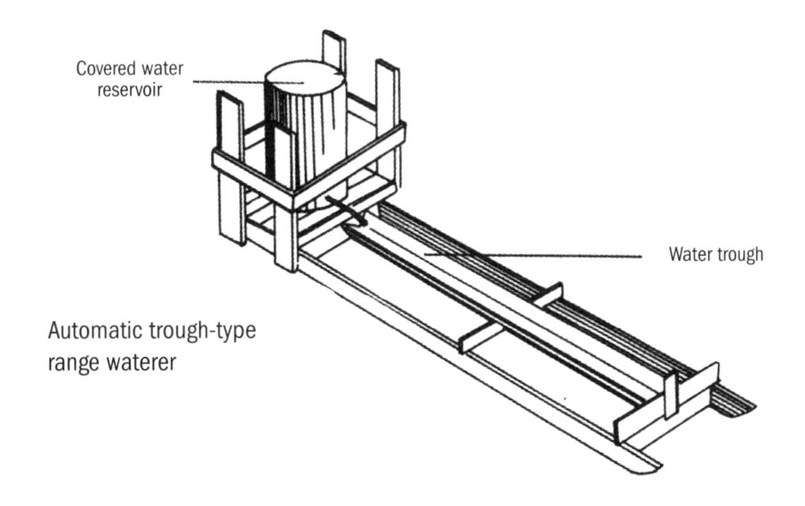

Automatic trough-type range waterer

Selecting a Range Site

Soil type and drainage. Poorly drained soil does not make good range for turkeys. Stagnant surface water can be a source of disease. Therefore, quick and complete drainage of turkey ranges is essential regardless of location. Range site selection in certain geographical areas may be dictated by soil type.

MATERIALS FOR TURKEY SUN PORCH	
5	1 × 4 common boards for door, 4' (1.2 m) long
8	2 × 4 corner posts, 5' (1.5 m) long
2	2 × 4 center studs, 5' (1.5 m) long
1	2 × 4 center stud, 14' (4.3 m) long
5	2 × 4 floor joists, 8' (2.4 m) long
2	2 × 4 floor joists, 12' (3.7 m) long
2	2 × 4 plates, 12' (3.7 m) long
2	2 × 4 plates, 8' (2.4 m) long
2	2 × 4 roosts, 8' (2.4 m) long
1	2 × 4 roost, 12' (3.7 m) long
1	2 × 8 rafter, 8' (2.4 m) long
1	2 × 4 at feeder, 6' (1.8 m) long
2	2 × 4 cross braces, 8' (2.4 m) long
33	1½ × 1½ fir boards for flooring, 12' (3.7 m) long (or 96 sq. ft. [8.9 sq m] of 1½ × 1½ turkey wire mesh)
26	1 × 8 common boards, D4S or T & G, 4' (1.2 m) long
14	1 × 8 common boards, D4S or T & G, 5' (1.5 m) long
4	1 × 12 common boards, D4S, 6' (1.8 m) long
1	1 × 12 common board, D4S, 8' (2.4 m) long
2	1 × 3 common boards, D4S, 6' (1.8 m) long
1	1 × 6 common board, D4S, 6' (1.8 m) long
1	½ × ½ strip, 6' (1.8 m) long
50	linear feet (15.2 m) of 2" (5.1 cm) lath for feed rack
50	linear feet (15.2 m) heavy galvanized wire, 8 or 9 gauge

Sandy soils are well suited for range-reared turkeys because they provide good drainage. On sandy soil, ranges that are flat or that have little slope can be utilized.

In areas with heavy clay soils, ranging on flat terrain is undesirable because of drainage problems. In such areas, a good ground cover for the open areas, such as fescue or orchard grass, helps both to stabilize the slope and to prevent muddy areas.

1	piece of ¼" (0.6 cm) tempered CDX plywood with exterior glue, 3' × 4' (0.9 m × 1.2 m)
3	3½" or 4" (8.9 cm or 10.2 cm) light T-hinges
3	4" (10.2 cm) hasps
3	padlocks
32	linear feet 2" (9.8 m, 5.1 cm) poultry wire, 48" (122.2 cm) high
12	linear feet 2" (3.7 m, 5.1 cm) mesh poultry wire, 36" (91.4 cm) high
1	roll roofing paper
5	pounds (2.3 kg) 6d nails — com
10	pounds (4.5 kg) 8d nails — com
10	pounds (4.5 kg) 10d nails — com
5	pounds (2.3 kg) 16d nails — com
2	pounds (1.1 kg) ¾" (2.1 cm) galvanized staples
1	pound (0.5 kg) 1" (2.5 cm) galvanized roofing nails

6d = 6-penny size
com = common nails with round heads
D4S = dressed/planed on four sides
hasp = hinged metal strap secured by staples and pin
joist = small beams laid horizontally to support floor
plate = horizontal timber carrying rafters for roof
rafter = sloping timber of roof
stud = upright
T & G = tongue and groove

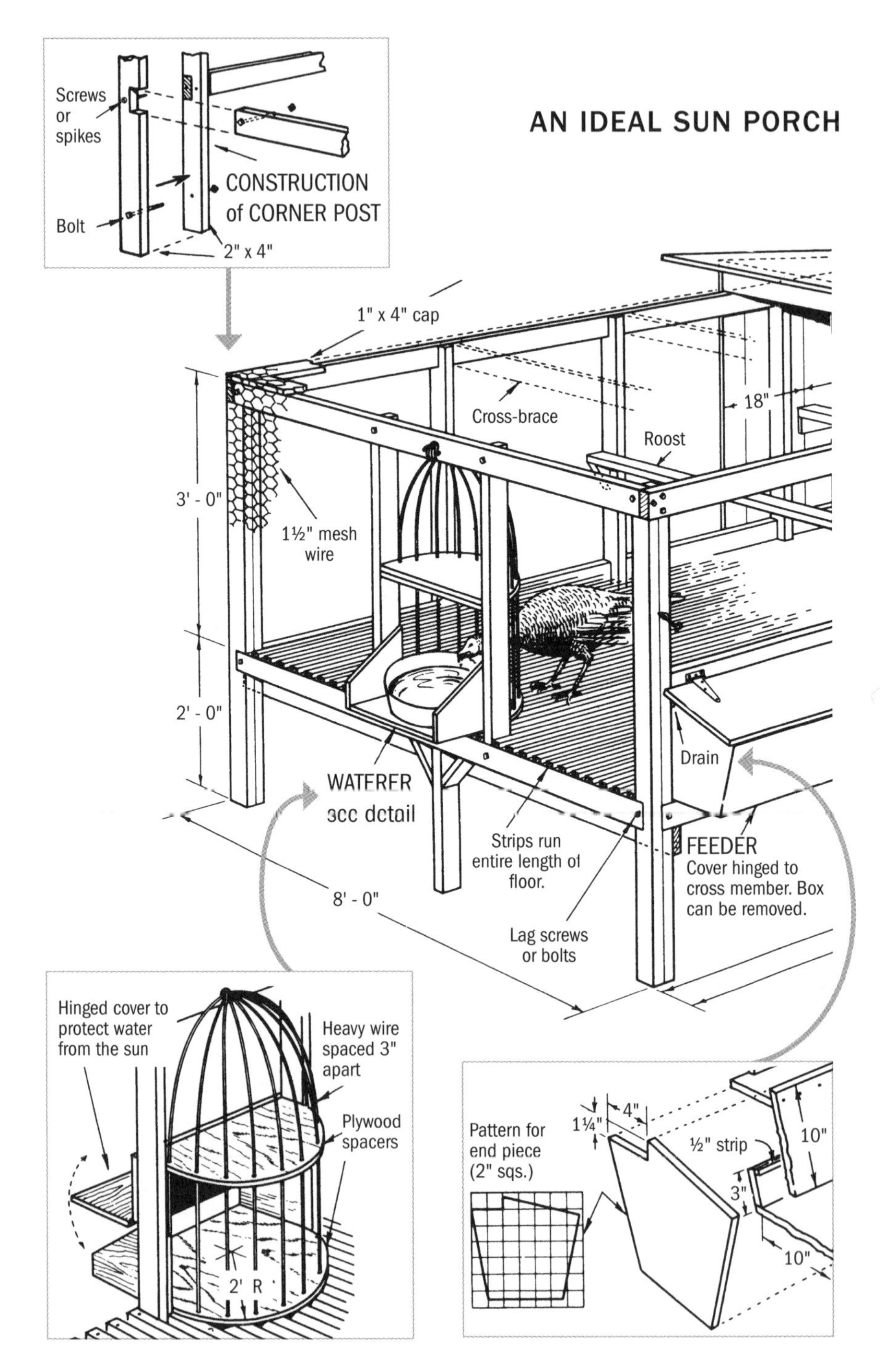

Plan and details for a house and sun porch system suitable for a small flock

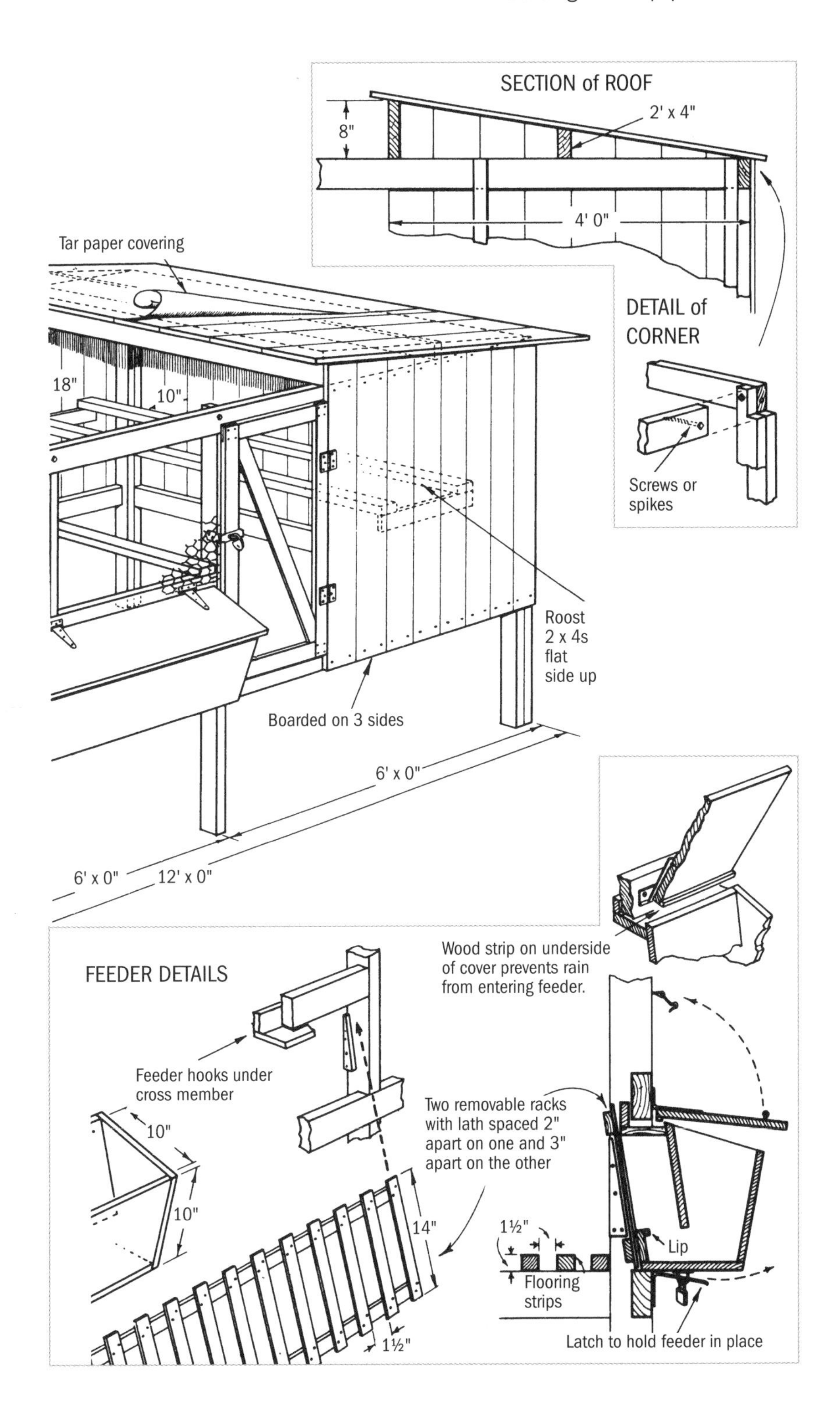
SECTION of ROOF
8"
2' x 4"
4' 0"
DETAIL of
CORNER
Screws or
spikes
Tar paper covering
18"
10"
Roost
2 x 4s
flat
side up
Boarded on 3 sides
6' x 0"
6' x 0"
12' x 0"
FEEDER DETAILS
Wood strip on underside
of cover prevents rain
from entering feeder.
Feeder hooks under
cross member
10"
10"
Two removable racks
with lath spaced 2"
apart on one and 3"
apart on the other
14"
1½"
1½"
Lip
Flooring
strips
Latch to hold feeder in place

Natural cover. The presence or absence of wooded growth or natural cover influences range site selection. Natural cover and shaded areas are important for turkeys reared on range. Ideally, the range should have both open and shaded areas. Partial shade is extremely important for turkeys reared during the hot summer months. Shade also helps reduce the energy required to lower body temperature and reduces incidents of heat stroke.

Confinement-Rearing

Rearing birds in confinement offers several advantages. It protects against losses from soilborne diseases, predators, thefts, and adverse weather conditions; labor costs and acreage requirements are less; and it reduces the direct effects of poultry on the environment.

Small-flock producers have numerous housing and management-system options. The brooder house or pen, if large enough, may be used to confine the birds until they reach market age. Several variations of confinement-rearing systems are used by small-flock owners — the house

MATERIALS FOR POULTRY HOUSE

Foundation	12 concrete blocks 8" × 8" × 16" (20.3 cm × 20.3 cm × 40.6 cm)
Floor joists	7 pc. 2 × 6 × 10' (0.6 × 1.8 × 3 m) lumber
Front and rear sills	2 pc. 2 × 6 × 12' (0.6 × 1.8 × 3.7 m) lumber
Floor	150 board feet (1,393.5 sq m) T & G sheathing
Shoe	54 linear feet (16.5 m) 2 x 4
Studs: Rear	9 pc. 2 × 4 × 5' (0.6 × 1.2 × 1.5 m)
Front	9 pc. 2 × 4 × 7' (0.6 × 1.2 × 2.1 m)
Ends	8 pc. 2 × 4 × 12' (0.6 × 1.2 × 3.7 m)
Partitions	2 pc. 2 × 4 × 12' (0.6 × 1.2 × 3.7 m)
Plates	2 pc. 2 × 4 × 12' (0.6 × 1.2 × 3.7 m)
Roof	175 board feet (1,625.8 sq m) T & G sheathing
	1½ squares of roll roofing

and porch system is one example. Some growers, after removing the birds from the brooding facilities, confine them to a wire-enclosed porch for added protection.

When birds are raised in strict confinement, adequate floor space is important. If the birds' beaks are trimmed, if feed and water space are adequate, and other conditions are optimal, large males can be confined to approximately 5 square feet (0.47 sq m) of floor space, females to 3 square feet (0.28 sq m), and mixed flocks to 4 square feet (0.37 sq m) per bird. Smaller varieties need 4 square feet (0.37 sq m) for males, 3 square feet (0.28 sq m) for females, and 3½ square feet (0.33 sq m) for mixed flocks.

The Yard System

When birds are reared in confinement, they need considerable floor space toward the end of the growing period. By using a yard attached to the housing facility, more birds can be kept in a smaller housing area. The yard should be well drained. You can put gravel or stones in the yard

Windows	2 pc. 2 × 10' (0.6 m × 3 m) fiberglass (flat)
Siding and doors	11 sheets of 4' × 8' × ½" (1.2 m × 2.4 m × 1.3 cm) exterior plywood
Miscellaneous	4 pc. 2 × 4 × 12' (0.6 × 1.2 × 3.7 m) lumber framing
Rafters	7 pc. 2 × 6 × 12' (0.6 × 1.8 × 3.7 m)
Anchors	6 pc. 1½" × ¼" × 12" (3.8 cm × 0.6 cm × 30.5 cm) steel
Fascia	2 pc. 1 × 6 × 12' (0.3 × 1.8 × 3.7 m)
Door stops	2 pc. 1 × 2 × 12' (0.3 × 0.6 × 3.7 m) 1 pc. 1 × 2 × 6' (0.3 × 0.6 × 1.8 m)
Nails and hardware	

T & G = tongue and groove / **o.c.** = on center

POULTRY HOUSE

Plan and elevations for a 10 × 12-foot (3.1 × 3.7 m) poultry house
(Note: Consult local health and building code authorities before starting construction.)

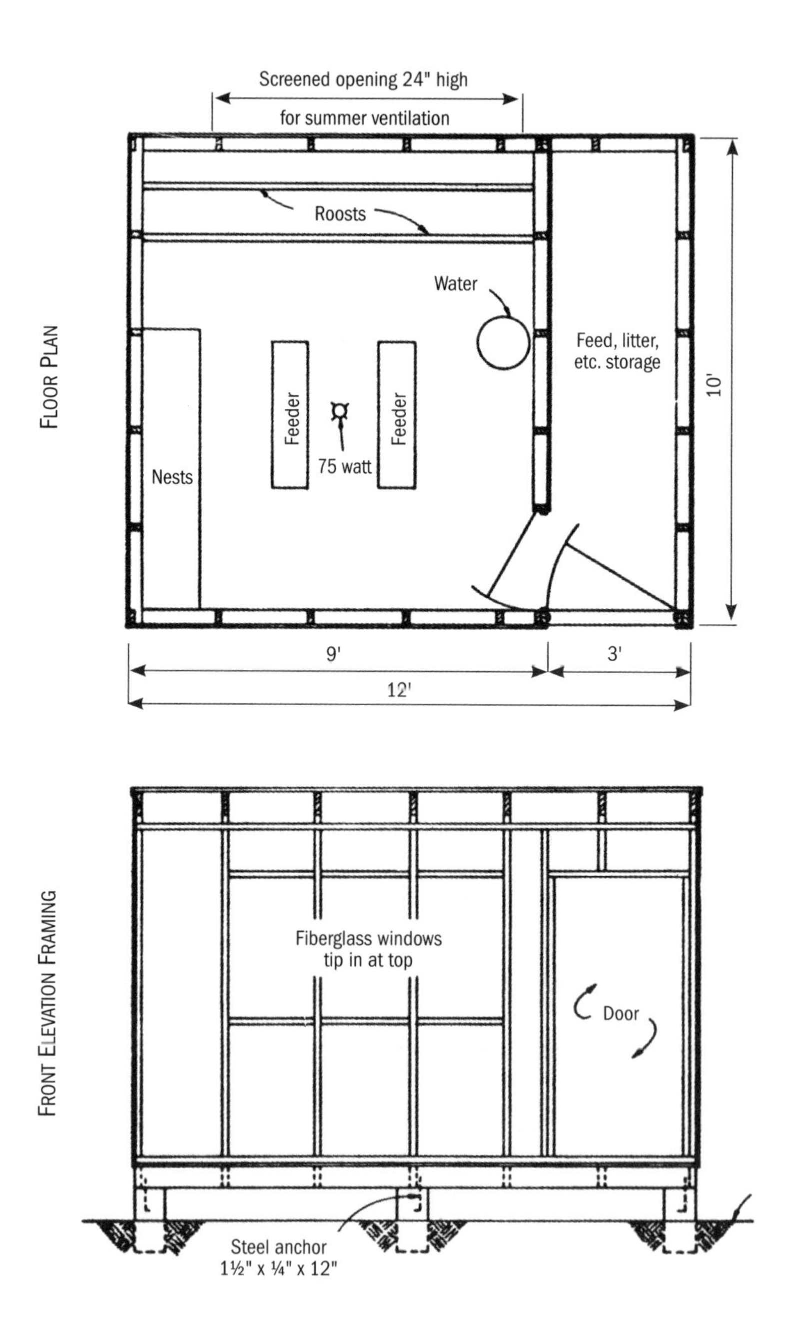

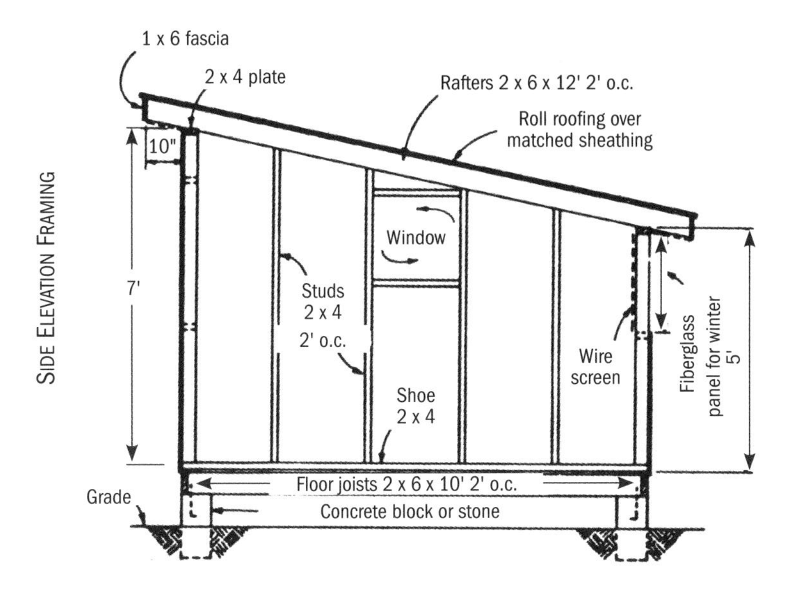

Framing for a 10 × 12-foot (3.1 × 3.7 m) poultry house

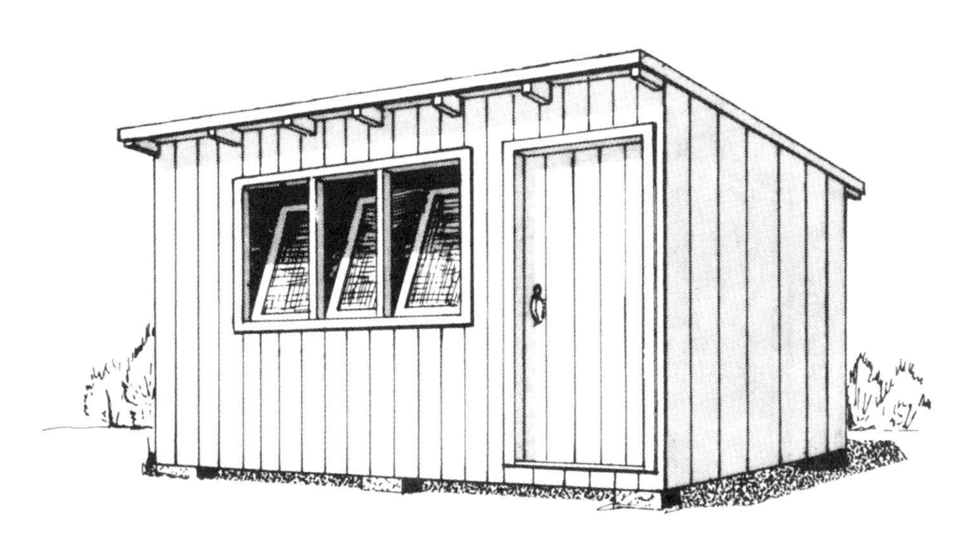

The completed poultry house can be easily ventilated and provides adequate floor space for the poults.

GUIDELINES FOR PROVIDING ADEQUATE FLOOR SPACE

It's important to provide adequate floor space for poults to avoid such problems as cannibalism. For heavy varieties, provide 1 square foot (0.09 sq m) of floor space per poult up to six weeks of age. From six to twelve weeks, increase the floor space to 2 square feet (0.18 sq m) per poult; from twelve to sixteen weeks, allow a minimum of 3 square feet (0.27 sq m). Mixed sexes grown in confinement need 4 square feet (0.37 sq m) of floor space per bird from sixteen weeks to market. If the flock is all toms, provide 4½ to 5 square feet (0.41–0.45 sq m) of floor space; if it is all hens, 3 square feet (0.27 sq m). For light-type turkeys, floor space requirements may be reduced slightly. Keep in mind that these space suggestions are the minimum necessary amount of space.

to keep the birds out of the mud, to improve sanitation, and to prevent disease. Yards need to be kept in good condition. Ideally, the location of the yard should be changed every 1 or 2 years. At least 4 to 5 square feet (0.37–0.45 sq m) of yard area per bird is recommended.

A woven-wire poultry fence 6 feet (1.8 m) high normally keeps the turkeys inside, but in some cases it may be necessary to clip the flight feathers or primary feathers on one wing (see page 120) to prevent the birds from flying over the fence. Lighting the range with floodlights also helps keep out unwanted animals and discourages raiding of the range or yard by predators. Electric fencing also serves this purpose.

Housing

If you start a small turkey flock in the warm months of the year, housing does not have to be fancy. However, the brooder house should be a reasonably well-constructed building that can be readily ventilated. If a small building is not available, perhaps you can provide a pen space within a larger building.

The brooder area should have good floors that can be easily cleaned and disinfected. Concrete floors are preferred, but wood floors are acceptable.

Insulation

The amount of insulation required in the building depends on the time of the year that the turkey poults are started, as well as climatic conditions in your area. A well-insulated building conserves energy, lowers brooding costs, helps keep the young turkeys warm and dry, and makes it possible to start turkey poults during any season of the year.

Lighting

Sunlight is not necessary for brooding turkeys. However, small-flock producers may use the building for the entire growth period; in this case, adequate ventilation and sunlight are essential. Windows must be placed to provide cross ventilation and necessary ventilation at critical times. Windows that tilt from the top and are equipped with antidraft shields on the sides provide good ventilation. It's important to be able to regulate windows and put them where drafts won't chill the young poults. One square foot (0.09 sq m) of window area per 10 feet (3 m) of floor space is normally adequate.

Equip the pen with electricity and artificial light. Young poults need intense light to enable them to find feed and water, thereby preventing starvation or dehydration. For the first 2 weeks, provide a minimum of 1000 to 1600 lumens of light at the poult level. Bright light should be used 24 hours a day for the first 3 days. A dim night-light is usually provided thereafter to prevent piling of the confined birds. Depending on environmental conditions, brooding of the poults is usually completed after 5 or 6 weeks.

Thoroughly clean and disinfect the area to be used for brooding poults. Many good disinfectants are approved for poultry and available from agriculture-supply houses. Whatever disinfectant is used, follow the directions on the container. Some disinfectants can cause injury to feet or eyes, which may damage the poults severely.

Equipment

The basic equipment required for growing a flock of turkeys includes brooders, feeders, and waterers. Breeder birds require additional items, such as laying nests and egg-handling equipment.

While turkeys require some special equipment for best rearing results, most pieces can be homemade or purchased from local feed and farm-supply outlets or by mail order (see resources). Nothing has to be fancy; however, it is important that feeding and watering equipment be designed to adequately service the birds with a minimum of spillage or waste.

The amount of equipment required, as well as its size, varies with the age and size of the turkeys. Sufficient feeder and water space must be provided to allow each bird equal access. (See also chapter 6.)

Brooding Equipment

Several types of brooders are suitable for poults. The heat source may be gas or electric. If a hover-type brooder (that is, one with a canopy over the heat source) is used, allow 12 to 13 square inches (77.4–83.9 sq cm) of hover or canopy space per poult. The brooder should be equipped with a thermometer that can be easily read. Take temperature readings at the edge of the hover approximately 2 inches (5.1 cm) above the floor. (See chapter 6 for more on brooder temperatures.)

The brooder size may be adjusted to meet your needs. For example, an 18 × 18-inch (45.7 × 45.7 cm) brooder would be suitable for 25 birds.

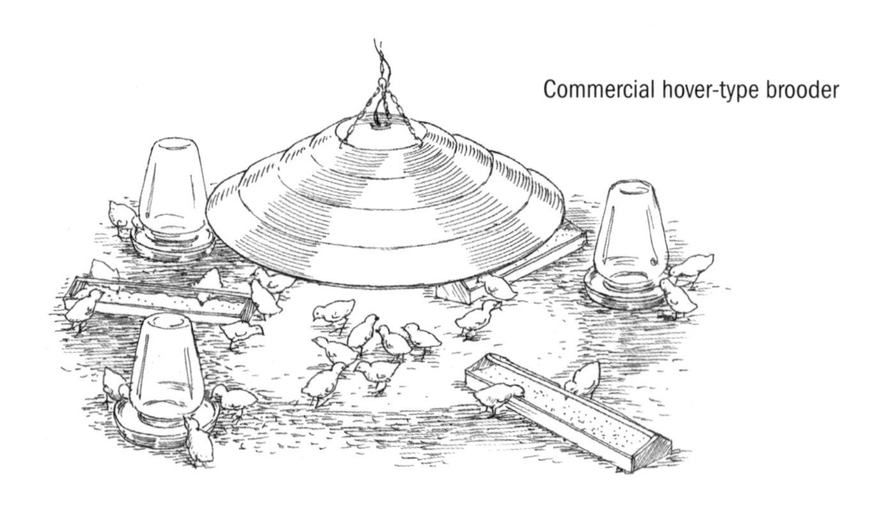

Commercial hover-type brooder

Infrared lamps are satisfactory for small numbers of poults. Provide two or three 250-watt bulbs per 100 poults. Even though one lamp may be adequate for the number of poults started, an additional bulb is recommended as a safety factor in case a bulb burns out. Hang infrared lamps about 18 inches (45.7 cm) from the surface of the litter at the start. After the first week, raise them 2 inches (5.1 cm) each week until they reach a height of 24 inches (61 cm) above the litter. The room temperature outside the hover or brooder area should be approximately 70°F (21°C) for maximum poult comfort.

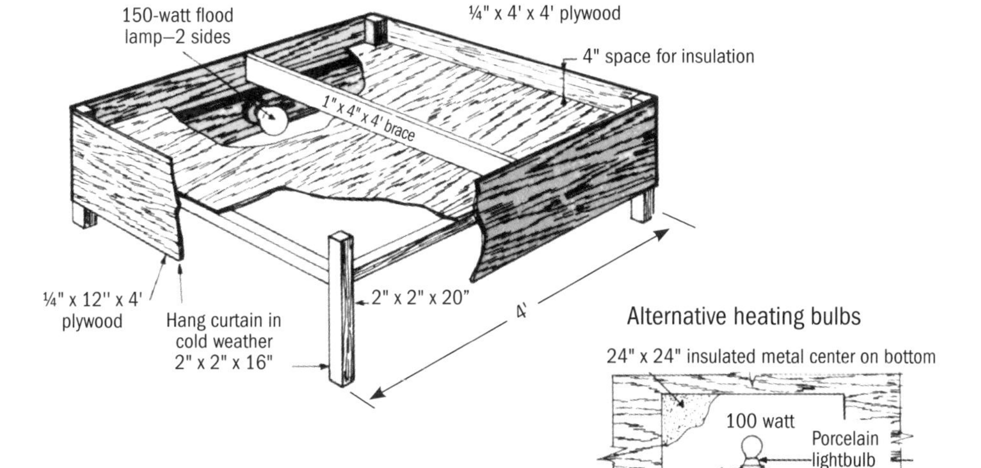

Battery brooder

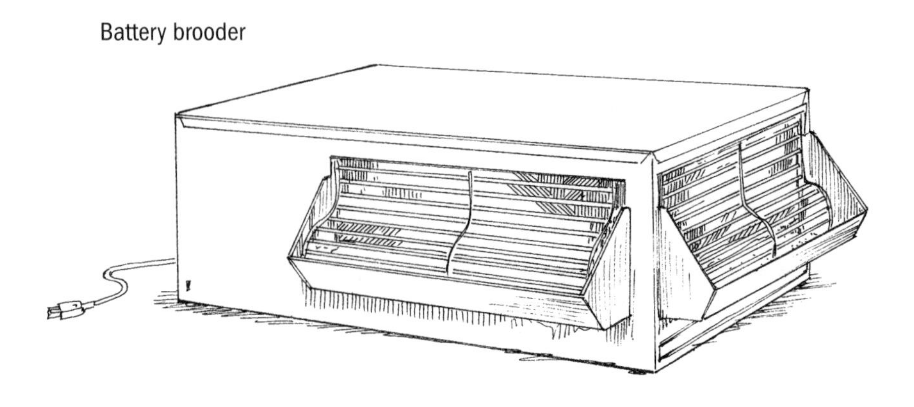

Where available, battery brooders can be used for the first 7 to 10 days to get the poults started. Allow approximately 25 square inches (161.3 sq cm) of space per poult in the battery. After you remove the poults from the battery and place them on the floor, watch carefully to make sure they learn to use the feeders and waterers. Also, poults that have been reared in battery brooders and then put onto floors with pine shavings as a bedding material will probably consume some of the shavings. This could lead to crop impaction. Providing a little grit in the battery brooder may help to develop the gizzard and prevent crop impaction.

Shallow pans or never used egg cartons are fine for beginning feeders.

Feeding Equipment

To quickly get the poults to start eating, place their first feed on egg-filler flats, chick box lids, paper plates, small plastic trays, or box covers. When one or more of the birds starts to peck at the feed, it will attract the others.

When box tops or egg-filler flats are used as the early feeders, spread them around in the brooding area among the regular-type feeders. Usually, at 7 to 10 days, the early feeders can be removed and the poults will use the regular feeders. You may want to lay paper underneath the feeders for the first few days to prevent litter eating. If you choose not to cover the litter with paper, be sure not to fill the feeders so full that the feed overflows onto the litter, because this may lead to the practice of litter eating. Use of smooth-surfaced paper is not recommended because slippery surfaces, used for prolonged periods, can cause foot and leg problems in young poults. If available, use paper with a rough surface.

Hanging, tube-type feeders and waterers are excellent for poults and for turkeys.

It is also a good idea to dip each poult's beak into a water trough to make sure the bird experiences where to find water. Dehydration is a common malaise of poults.

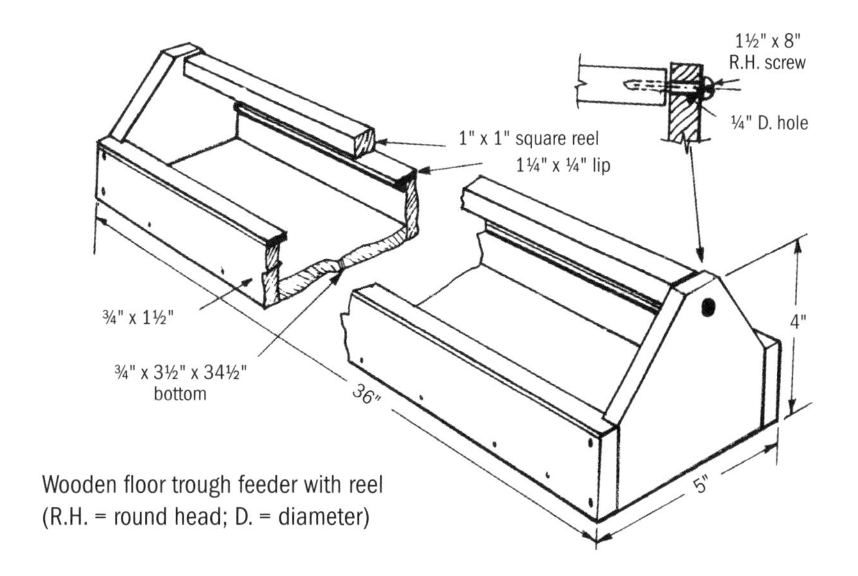

Wooden floor trough feeder with reel
(R.H. = round head; D. = diameter)

DETERMINING FEEDER SIZE

From 7 days to 3 to 6 weeks of age, use small feeders. Provide 2 linear inches (5.1 cm) of feeder space per bird. From 3 weeks to 12 weeks, the poults should have access to larger feeders about 4 inches (10.2 cm) deep, with 3 linear inches (7.6 cm) of feeder space per bird.

When figuring feeder space, remember to multiply the hopper length by 2 if the poults are able to use both sides of the feed hopper. Thus, a 4-foot (10.2 cm) trough feeder actually provides 8 linear feet (2.4 m) of feeder space. There are several types of feeders that can be purchased or built at home.

Hanging, tube-type feeders are excellent for turkey poults. The amount of available tube-type feeder space can be determined by multiplying the diameter of the feeder pan by 3.

After 7 to 10 days, the young poults should be acclimated to the location of the feeders and waterers and fully utilizing them. At that point, the paper can be removed and the poults allowed free access to the litter. Carefully observe the birds and how well they use the feeders and waterers to know when it is appropriate to remove the paper.

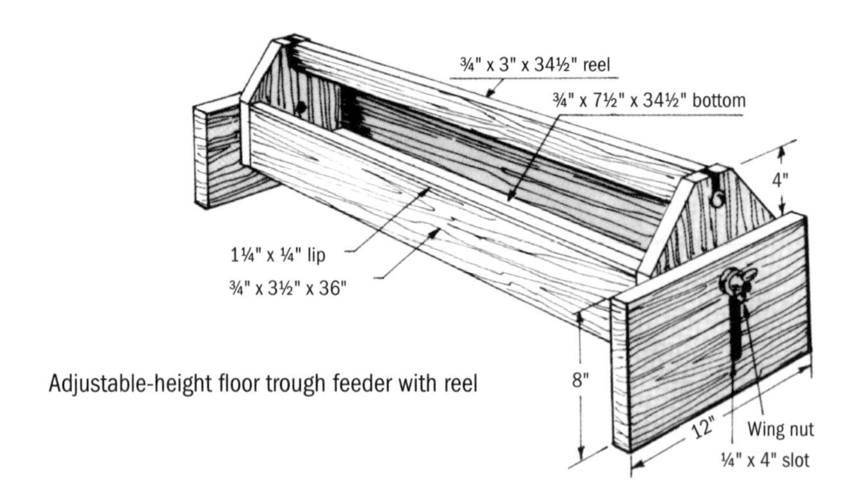

Adjustable-height floor trough feeder with reel

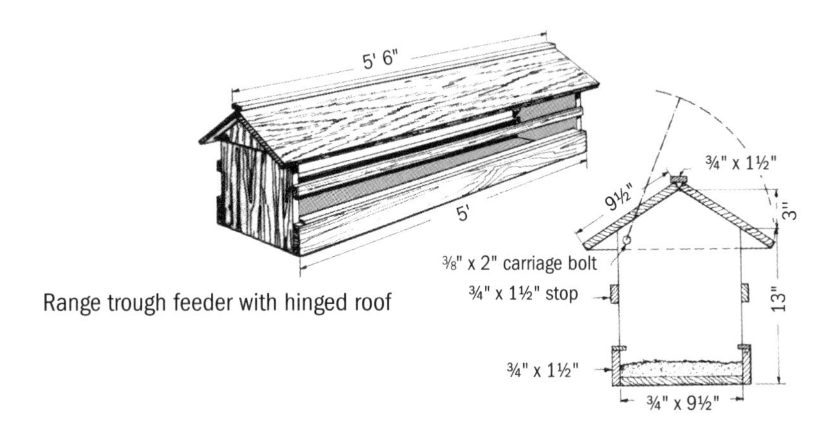

Range trough feeder with hinged roof

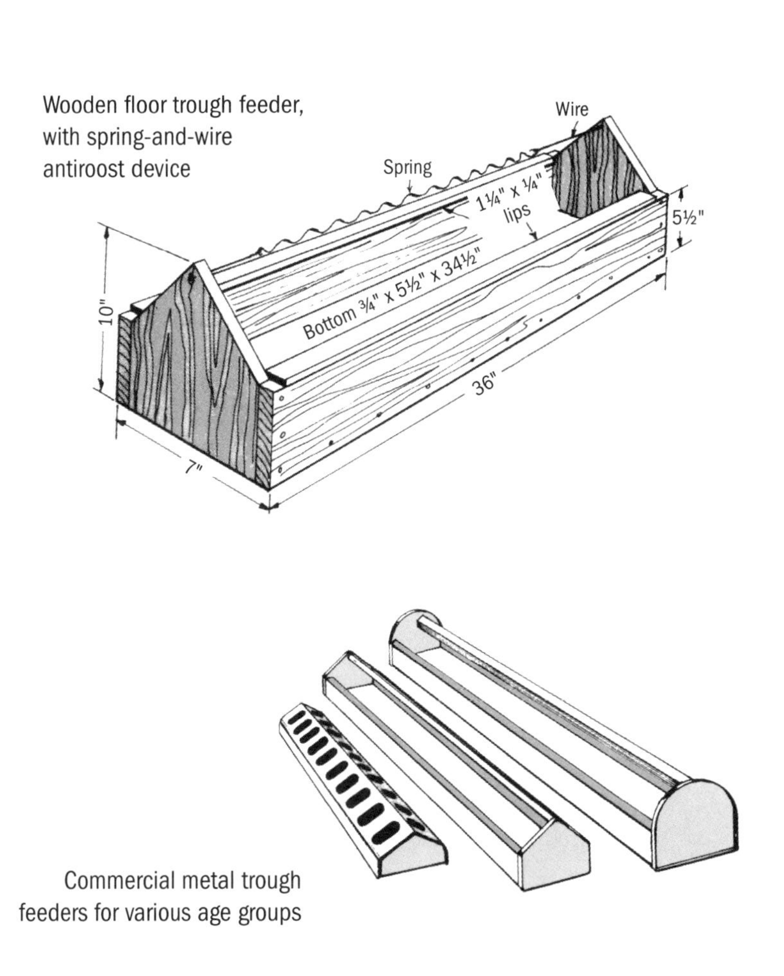

Wooden floor trough feeder, with spring-and-wire antiroost device

Commercial metal trough feeders for various age groups

Watering Equipment

Poults are usually started on either glass or plastic fountain-type waterers or automatic waterers. From one day to three weeks of age, they should have access to three 1- or 2-gallon (3.8 or 7.6 L) fountains per 100 poults. From three to twelve weeks, they should have two 5-gallon (19.2 L) fountains per 100 poults or one 4-foot (1.2 m) automatic waterer or two small bell-type waterers. For smaller flocks, adjust the number and size of waterers as necessary.

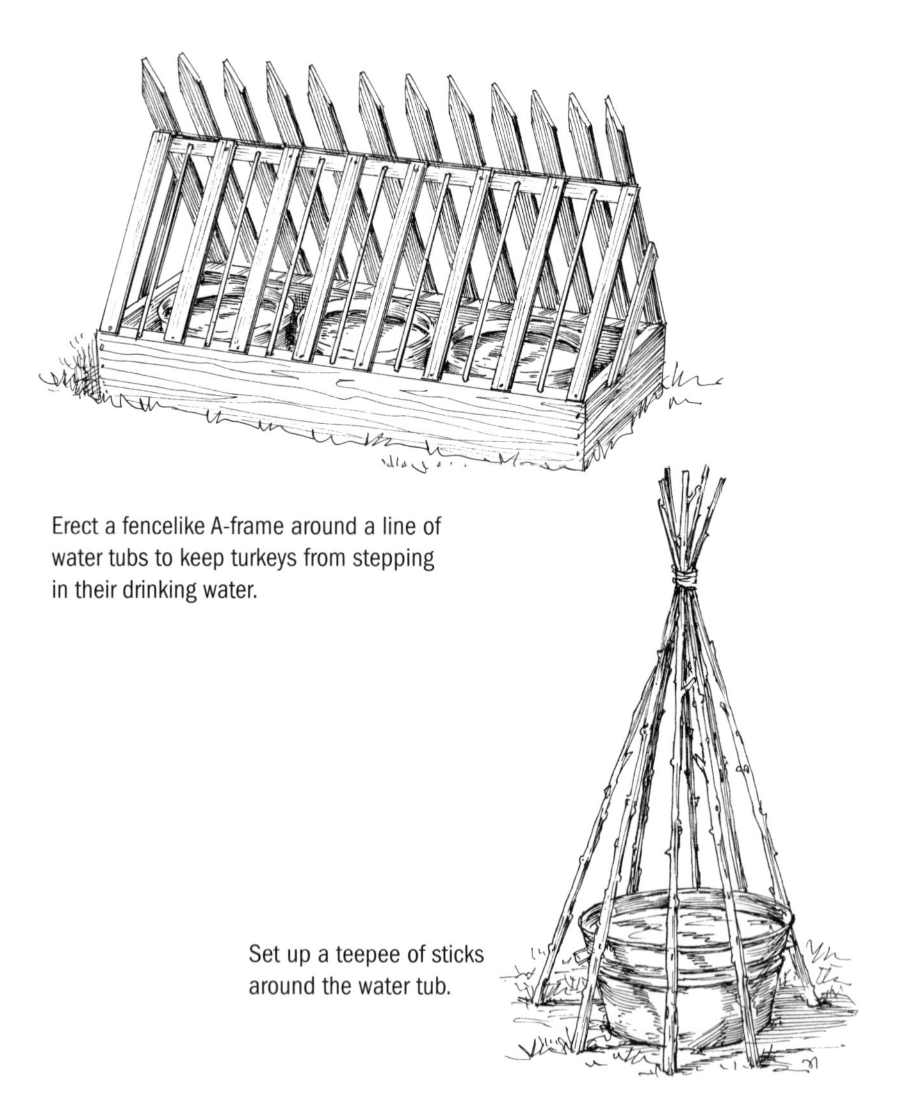

Erect a fencelike A-frame around a line of water tubs to keep turkeys from stepping in their drinking water.

Set up a teepee of sticks around the water tub.

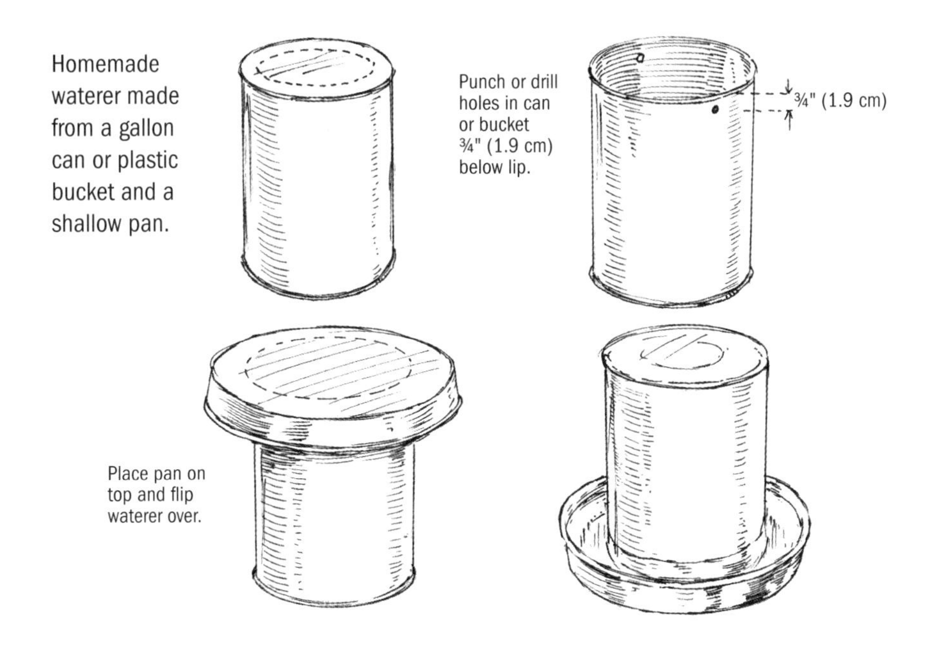

Note that providing adequate water space and water is imperative for good turkey production in any season but is especially important during warm and hot weather. A 20°F (11°C) increase in ambient temperature can double water consumption. Water temperature can also affect consumption. Providing water that is cooler than ambient temperature during hot weather is a good idea because it encourages the birds to drink and helps alleviate heat stress.

It is also imperative to keep water troughs clean. There is no magic number of times per week that waterers need to be cleaned — it takes as many times as it takes! You can find automatic waterers and water tanks at poultry-, livestock-, or agriculture-supply centers or online.

Miscellaneous Considerations for Feeding and Watering

Change equipment, both feeders and waterers, gradually to avoid discouraging feed and water consumption. For older birds, waterers can be placed on wire platforms. (The dimensions of the wire platform depend on the size and type of waterer used.) This helps prevent litter from fouling the waterers, keeps the poults out of the wet litter that frequently surrounds the waterers, and keeps the litter in better condition.

TYPES OF WATERERS

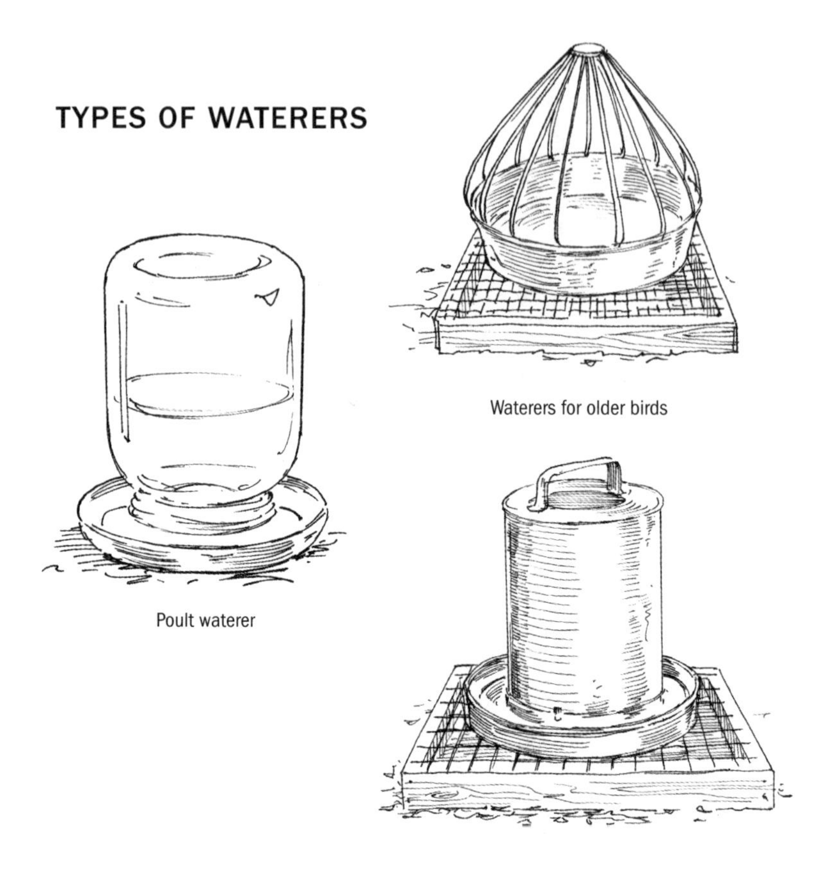

Waterers for older birds

Poult waterer

Alternatively, move the waterers often to avoid wet litter buildup. Once the waterer is moved, the litter beneath the old location can be turned to encourage drying, or it can be removed as needed.

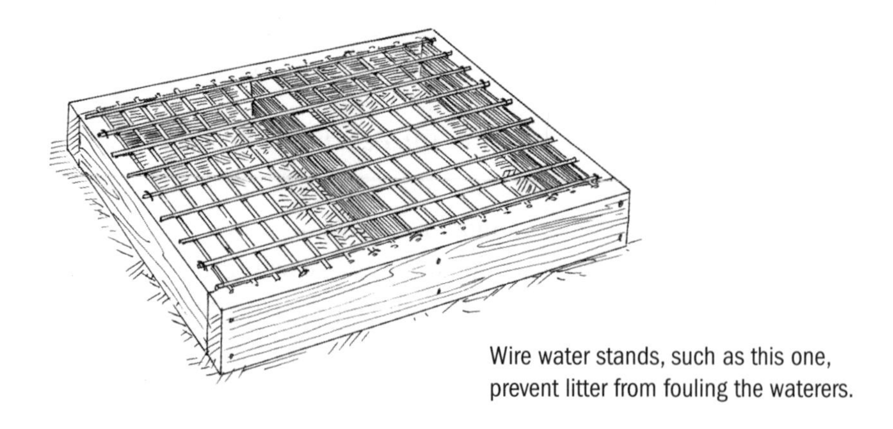

Wire water stands, such as this one, prevent litter from fouling the waterers.

Roosts

Roosts, or perches, are not used frequently for turkeys during the brooding period and are not necessary, though they do help prevent piling at night. Sometimes flashing lights, sudden noises, or rodents running across the floor can startle the poults. They may pile into a corner and cause injury or smothering. Normally, the birds begin to use roosts at four or five weeks of age. If roosts are used, they may be either the stepladder type or merely flat frames with perches on top. Following are some recommendations for roosts in the brooder house:

- Make roosts out of 2 × 4-inch (5–10 cm) boards turned so that the surface the turkeys will sit on is the 4-inch side.
- Place them 12 to 15 inches (30.5–38.1 cm) above the floor.
- Allow a minimum of 10 inches (25.4 cm) of roost space per poult by the end of the brooding period.
- Screen roost pits to keep the poults away from the droppings.
- If they are placed in a house or shelter, the roosts may be slanted to conserve space.

Where birds are grown on range, roosts are used quite frequently for young and smaller birds. Older birds, especially market-age toms, will probably not use roosts. Remove the roosts if space is limited and the birds have stopped using them.

Homebuilt roost for range turkeys

Roosts for range birds can be constructed in the following manner:

- Space the roosts 24 inches (61.6 cm) apart, 15 to 30 inches (38.1–76.2 cm) off the ground.
- Build outdoor roosts out of fairly heavy material to prevent breaking when the weight of the birds is concentrated in a small area.
- Allow 10 to 15 inches (25.4–38.1 cm) of perch space per bird for large-type birds and 10 to 12 inches (25.4–30.5 cm) for small-type turkeys up to maturity.

Litter Conditions

When turkeys are grown completely in confinement houses, roosts are normally not used. The birds bed down in the litter on the floor and, therefore, litter conditions must be sanitary to prevent such problems as breast blisters, soiled and matted feathers, and off-colored skin blemishes on the breast. More important, good litter conditions improve sanitation and prevent disease.

To maintain optimal litter conditions, remove wet or caked litter and replace with clean, dry material. Good pen ventilation helps remove excess moisture and keeps litter dry, as does rigorous management of water and waterers.

> *Do those things which you know ought to be done when they ought to be done. Turkeys wait for no one.*
>
> — Norman Kardosh, *master turkey breeder,* Alton, Kansas

4

Feeds and Feeding

PROPER NUTRITION FOR TURKEYS is a complicated business with many factors that can alter results: environment, exercise, and population density all affect feed performance. While wild turkeys range for 100 percent of their diet, and industrial turkeys are provided scientifically formulated complete rations, pasture-raised turkeys will consume both provided and wild foods. Watching and observing their growth and behaviors will help you properly feed your turkeys.

Feed Types

As wild relatives of today's farm turkeys forage throughout the course of the year, their diets vary according to seasonal availability. Wild turkeys eat bits of various plants, such as chickweed, grasses, and clovers, as well as insects — gnats, ants, leafhoppers, black field crickets, grasshoppers, slugs, small spiders, centipedes, various flies — and acorns and other seeds. These "wild" foods certainly can be a part of the farm turkey's diet, but farm birds require a larger volume of food than their wild cousins do as they grow to a larger weight and size.

Fortunately, almost any feed store sells premixed turkey feed nearly anywhere in America. Premixed feeds are formulated to satisfy the nutritional needs of turkeys; the formulas have been derived from scientific studies and refined over the past century. Feeds have been designed to satisfy the needs of turkeys at specific stages of life — the first few weeks (starter feed), the next 20 weeks (grower feed), for reproduction (breeder feeds), and during adulthood (conditioner or holder feeds).

Mash or Premixed Feed

Store-bought feed consists of many different ingredients. To prevent the turkeys from picking out favorite grains without consuming other needed nutrients, the feed is milled to a fine, flour-like consistency called "mash." The ground ingredients are mixed together evenly and consumed in appropriate proportions. In the early 1900s, mash was often mixed with water, milk, or whey and fed wet — producers called it "wet mash." Feeding wet mash has its drawbacks, leaving a mess that can attract flies or spoil if not consumed rapidly, but it is still an excellent way to encourage the birds to consume the remaining bits from their feed trough.

Mash became so common a method of mixing feeds that today we roughly refer to any premade feed as "mash" to differentiate such from other feeds, such as grains. Processed feed is now fed in several different forms; the most common are pellets, crumbles, flourlike mash, and combination concentrate-and-grain diets.

Pellets are formed from the flourlike ground mash. Bagged and sold as formed, these pellets make a feed form that is easy to digest, low in dust, and has little waste, as it is large enough for the birds to see even when spilled on the ground. Pellets are too large for young poults.

Crumbles are crushed pellets. This feed form is a nice balance — holding mash ingredients together, as with pellets, with only a medium amount of waste (better than mash, worse than pellets). Crumbles are small enough for day-old poults to easily consume.

Combination concentrate-and-grain feed uses local grains and a pelleted concentrate that has all the minerals and vitamin supplements necessary for the feed to reach a particular level of nutrition. The advantage of combination concentrate-and-grain feed is that farm ingredients may be used — feed may be mixed on-farm or locally — and the turkeys receive balanced nutrition. The disadvantage is that it is possible for one turkey to eat too much of the concentrate, and another too little.

Premixed feed is readily available but has three drawbacks: all formulations assume the feed as the sole source of nutrition; the manufacturing of premixed feeds often utilizes heat and pressure, resulting in the breakdown of some nutrients; and processed feed has a short shelf life, becoming stale in the presence of high humidity. Poultry people who range their birds must understand that as the birds leave the brooder and begin to forage, dietary levels will change, dropping with the input of lower-protein foodstuffs such as grasses or rising with the consumption of protein-rich insects. Give balance by offering premixed feed; purchase feed often and store it properly so that it remains fresh and nutritious.

So how does one balance the diet of pastured birds? Let the birds balance their own diets.

Natural Balancing

Given the opportunity, poultry on pasture will balance their diets naturally, as long as they can access a variety of foodstuffs. Foods found in the wild provided life for the farm turkey's ancestors and are still appropriate for today's turkeys—as long as they can find enough to satisfy the volume their bodies need.

Base Feed

Poultry on pasture require a consistent base feed. This gives a source of balanced nutrition that the turkeys can consume in larger or smaller quantities, as desired. It lends consistency to their diets as seasons change and other foods become available. This is important as it reduces stress and keeps the beneficial gut bacteria constant. The gut bacteria that help digest food come in a variety of forms, each favoring certain food types. An ever-changing diet causes an ever-changing population of gut bacteria.

Bacteria take time to increase their population. As the diet changes, the bacteria designed to consume the new diet may have a small population, delaying their ability to pull nutrition from the diet. By providing a consistent base diet, the forms of gut bacteria that favor that diet remain as a strong population, and support the body consistently.

As turkeys go about finding food and balancing their diets, consider how a body goes from deficiency of a nutrient to a proper level;

"nutrient loading" describes the process of a body overcoming nutrient deficiency. The process is simple: consume the nutrient in greater quantity than the body utilizes it. It usually takes about 30 days to reach saturation — the point at which the body has stored all it needs and can now output the nutrient in its products such as eggs or manure. Once the nutrient level reaches saturation, the body needs to consume less to maintain this proper level. The experience is similar to having a craving and consuming quantities of the craved food over several days, and then becoming satiated so that the food is no longer desirable.

Following natural cravings, and then loading nutrients, is a way turkeys seek to balance their diets. Another action that leads to consumption at levels that satisfy the needs of a body is exercise.

Exercise

Exercise revs up most of the body's systems. Movement calls upon the muscles to work, and thus burn more calories to function, grow, and heal. The respiratory system must move more oxygen in and out of the creature. Exercise calls upon the circulatory system to deliver oxygen and nutrients at a faster pace. The digestive system must step up to provide the calorics used to operate all the systems at this pace. In simple terms, this means that as turkeys move around more, they eat more and grow faster.

When a turkey exercises regularly, it consumes a larger volume of feed and thus has more nutrients available. It can assimilate more of the nutrients it needs and expel any overabundance of other nutrients. Exercise is therefore more important than the exact feed provided.

Feeding

Turkeys need a high level of protein to support the rapid rate of growth they experience when they are young. High protein is especially necessary during the early half of their grow-out; less protein is necessary as they mature. During the first 4 to 6 weeks, growing poults should eat only mash, with a bit of some finely cracked corn to attract the young poults to the feeders.

Turkey poults need feed with a protein ratio of 20 to 30 percent in order to support their growth. It is a good practice to feed young poults a feed that contains 28 percent protein, if available. The protein levels of feeds vary, based on the mixes sold locally. Since protein is usually the most expensive ingredient in poultry feed, higher-protein feeds do cost more. Many people will opt to purchase a lower-protein feed at lower cost, but in many cases this is the more expensive plan to follow.

Various turkey feeders. You can purchase galvanized and plastic hanging feeders (top) from poultry equipment suppliers. You can also get kits to turn a 5-gallon plastic bucket (bottom left) into an inexpensive feeder, or cut a plastic drum into a great homemade feeder.

A good rule of thumb is "protein equals growth." If I restrict the protein level (feed a low-protein diet) during the first 10 to 16 weeks of a poult's life, that poult will never grow as fast or reach the same adult size that it could on better feed. Understand that this early period is when the skeleton and all the major organs receive the nutrition they need to grow to full potential.

A better plan is to feed high levels of protein early on and reduce the protein level as the birds mature. One way of doing this is to keep the turkeys on the same high-protein feed from brooder through grow-out and processing in November, but supplement their diet with grain as they mature. In this plan, a separate feeder is used to feed grain once the poults are placed on pasture. The separate feeder prevents feed waste from the poults using their beaks to scoop out the mash in order to find the grains.

Anytime around four to six weeks of age, begin to sprinkle a significant amount of grain over the top of the mash. From this time on, turkeys will eat more grain and less mash. Additional feeds may be used to supplement, such as green feeds and milk products, but mash and grain should be available to the turkeys at all times. Be sure to sprinkle some insoluble granite grit over the turkeys' feed a few times per week. The birds need this grit to grind the grains, and thus better metabolize them.

When grains are included as a part of the diet, the amount fed depends on the protein content of the mash or pellets the birds receive.

GIVE ME SPACE

When planning for turkeys, remember that adolescent and adult turkeys need 6 inches (15 cm) of space per bird at the feeder. That means that a two-sided, 5-foot-long (1.5 m) feeder will serve up to 20 turkeys. You can purchase or build feeders. Homemade feeders can be made from a variety of materials, including wooden boards or even PVC pipe, cut to produce a feed trough. Turkeys care little about the exact material used; however, they may shun feeders made of treated lumber.

Remember that grains contain low levels of protein (corn has 8 to 9 percent, wheat has 10 to 12 percent, and oats have 11 to 12 percent). Feeding birds too much of these grains dilutes the total protein content of the combined diet to the extent that it could affect growth. When the birds are twelve to sixteen weeks old, they can receive a grain and mash diet, but don't dilute the protein content below the 16 percent level. When feeding grains with the finishing diet, avoid diluting the diet with grains to the extent that total protein intake drops to less than 14 percent.

Offer poults grain in a feeder free choice, restricting the amount so they receive no more than one-fourth of their diet in grain until the age of sixteen weeks. This grain limitation will ensure they receive the protein they need — aim for a target of 16 percent protein or more. From the age of sixteen weeks on, increase the amount of grain being fed until young turkeys are eating largely grain (maybe three parts grain to one part mash) by October. Since grain costs less than high-protein mash, this plan reduces cost while nutrition supports the rate of growth. It also benefits the poults to be fed the same feed all along as the grain supplement gradually increases — this eliminates the stress of a changing diet.

Feed Management: The Key to Turkey Health

Feed quality is extremely important to the health of your turkeys. You can optimize performance by following these guidelines:

- Heat destroys feed nutrients, especially certain vitamins; store feed in a cool, dry area.
- Use a "batch" or "lot" of feed within 4 weeks of mixing, especially during the summer. Plan the amount of feed, mixed or purchased, based on your birds' rate of consumption.
- Monitor the quality of feed ingredients according to how they are handled and stored as well as their nutrient content. The final feed product is only as good as the ingredients used to mix the diet.
- Have your feed and feed ingredients analyzed at a laboratory on a regular basis. This is the best way to monitor feed quality.

By the time the effects of poor feed quality are demonstrated in bird performance, it is likely too late for the birds on that feed, and substandard performance will probably be the result. At the very best, you will lose time and profits by having to hold the birds longer than desired to overcome the effects of poor feed.

Feed Ingredients

Many different feeds have been used historically to feed turkeys on small farms. If you are trying to find methods that utilize as much feed from your own farm as possible, it is important to allow the turkeys a lot of range. Ranging turkeys glean much of their food in the form of insects, grubs, nuts, berries, and some green forage. Feeds you grow for your turkeys should consist largely of grains — which have the advantage of being easily stored. Protein is necessary, however, to balance the diet.

A method that was popular in the early 1900s was to add buttermilk to the turkey diet. The positives of this feed were a superb source of calcium, live-culture probiotics, and an excellent source of protein. The negatives included diarrhea from overfeeding, especially for poults one to four weeks of age; spoilage of mash when fed as wet mash; spoilage in water founts, especially during warm months, when fed as a drink; and the need for daily cleaning of the founts.

Other protein sources from the farm include alfalfa hay, field peas, and even soybeans (which must be roasted to destroy naturally occurring toxins). Protein is most important through the first 16 weeks of age; thereafter, turkeys digest carbohydrates more efficiently to produce gains. (The Missouri Experiment Station documented this protein–carbohydrate balance in 1943 while comparing growth rates of male versus female turkeys.)

Feed grains in a different hopper to prevent the turkeys from billing out and wasting the mash. A good rule of thumb is to begin fattening turkeys in October by feeding mostly grains.

Corn is an excellent turkey feed and produces a creamy yellow carcass. Wheat, barley, and grain sorghums are about equal to corn in feed value. Oats are a favorite turkey feed. Green feeds for turkeys include alfalfa, clovers, rape, Sudan grass, soybeans, Ladino clover, and vetch.

FOOD PLAN

Adult toms and hens will eat from 180 to 220 pounds (81.6–100 kg) of feed per bird per year.

APPROXIMATE PERCENTAGES OF INGREDIENTS FOR TYPICAL RATIONS

Ingredient	Starter	Grower 1	Grower 2	Finisher
Corn	44.30	56.00	63.20	66.50
Soybean meal (48%)	40.00	28.00	19.00	15.00
Fish meal	8.00	7.00	8.00	8.00
Fat	2.50	4.00	6.00	7.00
Dicalcium phosphate	2.40	2.20	1.60	1.50
Calcium carbonate	1.35	1.35	1.10	1.00
Salt	0.35	0.35	0.35	0.35
Choline chloride	0.20	0.10	–	–
Lysine	0.28	0.40	0.30	0.30
Methionine	0.25	0.18	0.15	0.10
Vitamin premix*	0.20	0.20	0.20	0.20
Mineral premix*	0.10	0.10	0.10	0.10
Coccidiostat*	0.07	0.07	–	–
Total	**100**	**100**	**100**	**100**
Calculated Approximate Analysis				
Crude protein	28.00	22.00	19.00	17.00
Metabolizable energy (kcal/lb)	1,300.00	1,380.00	1,475.00	1,510.00
Calcium	1.40	1.30	1.10	1.05
Available phosphorus	0.74	0.67	0.56	0.54
Methionine	0.71	0.56	0.50	0.42
Lysine	1.81	1.53	1.20	1.09
Sodium	0.19	0.18	0.18	0.18*

*Use at manufacturer's recommended rate.

Feeding Philosophy

In feeding turkeys, you must balance between providing them with enough nutrition to support their growth rate on the one hand, and encouraging them to make good use of forage on the other hand.

You can train turkeys to make good use of forage and other wild foods. It starts in the brooder, where poults receive daily opportunities to eat a variety of plant life while just a few days old. By the time the birds are on pasture, you support this learning by letting them completely utilize one area in a short period of time, and then offering them a new area. Turkeys take to this new stimulation and become accustomed to the rhythm of pasture rotation.

While brooding, feed tender alfalfa, white Dutch clover, tender young grass, or green grain sprouts, all chopped into short lengths and fed once or twice daily. Turkeys like tender green feeds such as short, fresh, unsprayed lawn clippings and garden vegetables such as Swiss chard, lettuce, and even the outer leaves of cabbage. Do not let the birds eat wilted, dry, or long, stringy roughage, as this type of feed can cause impacted or pendulous crops. Again, when feeding roughage, make sure the birds receive an insoluble grit, such as turkey-size granite grit.

Utilize the turkey's nature to keep feed costs down and birds healthy, happy, and active. Turkeys come off the roost hungry each morning; they have just spent one-third of the day without eating. If they find no mash first thing in the morning, they will work the pasture to find their breakfast. This is a good thing, and you should try to encourage this early-morning foraging as the turkeys benefit most from self-harvesting the pasture in this way. Make sure there are pasture plants and insects available, or the turkeys may turn on each other out of hunger and boredom.

It is a good practice to feed the turkeys at midmorning. Some folks like to give them all they can clean up in about 30 minutes, and then feed again late in the afternoon, all they can clean up. This works well if there is enough feed so that each turkey gets its fill. But if the turkeys do not get enough to eat, they will not finish at their full potential and will need more time to flesh out.

Most of us think of breakfast as the most important meal of the day. For turkeys on pasture, the most important meal of the day is the one they eat just before roosting. A turkey on good pasture can find its own breakfast. But a turkey on the roost at night has in its crop only what it could consume before roosting. Since night represents one-third of a turkey's life, the birds must be encouraged to fill up before roosting so they will not be hungry all night. A full crop gives them calories to burn to stay warm and to grow. If the grain hopper is filled with corn, close it off during the day and open in late afternoon to encourage the turkeys to eat a little bit more before roosting.

You can have some success feeding a premixed feed and simply keeping the feeders full, but in doing so you are squandering all the free, self-harvested forage your land has to offer. By working with the nature of turkeys, you can better assimilate them into the ecosystem of the farm and have happier, healthier turkeys as a result.

Feed Consumption

Turkeys are highly efficient at utilizing grains. The average heritage turkey will grow to market weight in 28 weeks and will have consumed 4 to 4.75 pounds (1.8–2 kg) of grains for each 1 pound (0.5 kg) of body weight gained, when raised on range. Compare this ratio change to a confined bird's consumption of 5.5 pounds (2.5 kg) of grain for every 1 pound (0.5 kg) of body weight gained. Of course, this is an average amount. The first few weeks of their lives, turkeys consume less and in the last few weeks before finishing they consume much more. Quality of feed and range conditions have a large impact on actual amount of feed consumed.

The 1945 book *Turkey Management* by Marsden and Martin reported that the Oregon Experiment Station found pasture advantageous and efficient in tom turkey production. Reductions in hen production are outweighed by gains by the toms in mixed-sex flocks.

FEED CONSUMED PER POUND OF TURKEY

Average Dressed Weights

Lot	Amount of Feed	Tom	Hen
1. No greens/sandlot	6.64 lbs (3 kg)	20.40 lbs (9.25 kg)	13.00 lbs (5.9 kg)
2. Dried greens/sandlot	6.50 lbs (2.9 kg)	22.90 lbs (10.4 kg)	13.30 lbs (6 kg)
3. Freshly cut alfalfa dryland yard	5.45 lbs (2.5 kg)	23.10 lbs (10.5 kg)	12.85 lbs (5.82 kg)
4. Alfalfa range	5.37 lbs (2.4 kg)	24.30 lbs (11 kg)	12.80 lbs (5.8 kg)

Lot 2 had 10% dried alfalfa leaves added to the feed. Lot 3 was fed all the freshly chopped alfalfa they could eat twice a day. Lot 4 was given access to a yard with alfalfa growing in it. All lots had all they could eat of turkey grower mash.

STOCKING DENSITY EFFECT ON FEED CONVERSION

Pounds of Feed Consumed Per Pound of Gain Per Acre

Unlimited Range	> 30 Turkeys	31–99 Turkeys	100+ Turkeys
4.5 lbs	4.8 lbs	5.9 lbs	6.1 lbs

Source: Marsden and Martin, *Turkey Management*. Marion Clawson's report on the impact of turkeys per acre on feed efficiency was based on a USDA survey completed by turkey producers from various areas of North America.

FEED CONSUMPTION PER BIRD

Pounds of feed consumed per bird during a 2-week period comparing heritage and broad-breasted turkeys

Age in Weeks	Heritage	Broad-Breasted
1–2	0.6 (0.27 kg)	0.6 (0.27 kg)
3–4	1.75 (0.8 kg)	1.75 (0.8 kg)
5–6	1.9 (0.86 kg)	3.6 (1.6 kg)
7–8	2.84 (1.3 kg)	5.5 (2.5 kg)
9–10	3.54 (1.6 kg)	7.75 (3.5 kg)
11–12	4.32 (1.9 kg)	9.8 (4.4 kg)
13–14	5.48 (2.48 kg)	11.9 (5.4 kg)
15–16	5.68 (2.6 kg)	12.5 (5.6 kg)
17–18	6.65 (3 kg)	15 (6.8 kg)
19–20	6.92 (3.1 kg)	15 (6.8 kg)
21–22	7.93 (3.6 kg)	–
23–24	8.24 (3.7 kg)	–
25–26	8.37 (3.8 kg)	–
27–28	8.71 (3.9 kg)	–

Source: The American Livestock Breeds Conservancy, *How to Raise Heritage Turkeys on Pasture*.

Turkeys on range still need a significant amount of feed. You can lower feed consumption by approximately 20 to 30 percent by utilizing quality range. The birds will also grow a little faster — the addition of exercise speeds up their metabolisms.

ATHLETES INDEED

Wild turkeys are strong fliers and have been known to fly up to speeds of 55 miles per hour over short distances.

If you are thinking of running a wild turkey down, think again. A wild turkey can run 15 to 30 miles per hour. Better put on your best running shoes!

Feeding the Breeders

Four weeks before the onset of egg production, start breeder turkeys on a turkey breeder diet. Studies have shown that it takes approximately three weeks for the effects of a quality feed to appear in the nutrition of the eggs; it takes a month for the full effect to appear. Since the young poult is a product of the egg's ingredients, the parent stock's feed is important; high-quality feed gives the poult an advantage during the first few days of life.

The turkey breeder diet usually provides 16 to 18 percent protein and 2.5 to 3 percent calcium. Complete pelleted diets are preferable, but protein concentrate and grain diets will work if fed in the right proportions. Hens must receive adequate calcium for proper eggshell formation. If hens eat too much grain in relation to the complete feed or concentrate, egg production and hatchability decrease. Provide hens with free-choice oyster shell to supplement the calcium in the feed. During the winter or any other rest period, hens should eat a holder (maintenance) ration, which has less protein and calcium than a breeder ration.

FEED CONSUMPTION OF TURKEY BREEDERS

Per Bird per Day

Type of Turkey	Toms	Hens
Large	1.5 lbs (0.68 kg)	0.8 lb (0.36 kg)
Medium	1.25 lbs (0.56 kg)	0.6 lb (0.27 kg)
Small	0.75 lb (0.34 kg)	0.5 lb (0.22 kg)

Source: Turkey Production, Agriculture Handbook No. 393, United States Department of Agriculture

A protein supplement and grain system has the advantage of using homegrown grains without the need for grinding or mixing. Homegrown grains can be ground and mixed with concentrate or custom-milled by the local feed store.

PERCENTAGES OF INGREDIENTS FOR TYPICAL BREEDER TURKEY RATIONS

Ingredient	Starter	Grower 1	Grower 2	Finisher
Corn	64.1	77.80	92.00	73.00
Soybean meal (48%)	22.5	11.50	–	14.50
Fat	4.0	2.00	2.00	2.00
Alfalfa meal or wheat midds	–	5.00	3.00	5.00
Calcium carbonate	6.5	1.00	–	2.75
Dicalcium phosphate	2.2	1.80	2.00	2.40
Salt	0.30	0.30	0.30	0.30
Methionine	0.10	0.25	0.25	–
Vitamin premix	0.20	0.20	0.20	0.20
Mineral premix	0.10	0.10	0.10	0.10
Total	**100**	**100**	**100**	**100**
Calculated Approximate Analysis				
Crude protein	16.00	12.50	8.00	14.00
Metabolizable energy (kcal/lb)	1,350.00	1,420.00	1,480.00	1,330.00
Calcium	3.0	0.9	0.47	1.6
Available phosphorus	0.5	0.4	0.46	0.6

Percentages are rounded off.

Feeding Breeder Toms

Breeder toms of heavy, commercial strains can become quite large and difficult to handle. It may be useful to restrict the amount of feed these birds receive. This is easy if toms are kept separate from hens. Once the toms are mature and are producing semen, feed them on a daily basis to maintain their weight or to allow them to gain slowly by using the restricted-fed ration given above. Alternatively, once the toms are producing semen, they can eat a pelleted diet made from corn with vitamins and minerals. This provides plenty of energy and protein for maintenance, but prevents rapid growth.

Breeder toms may eat the same ration as breeder hens, but take care to observe their body condition, as fat toms will not breed. Appraise body condition by pinching the pelvic bones, just below the vent on the abdomen of the bird. The thickness you feel here is comprised of skin, gristle, fat, and bone. A turkey in good condition will have about a half-inch of thickness over each bone. A turkey with more thickness is not likely to breed (if a tom) or lay (if a hen).

◆ ◆ ◆

Turkeys are efficient grain eaters and avid foragers. As they mature, their diets can include a great many foodstuffs produced on the farm. Keep in mind that different turkey types require different volumes of nutrition. Wild turkeys will harvest, given the opportunity and enough land to range, all of their diet; heritage/Standard-bred turkeys will harvest a significant portion of their diet and consume farm-produced grains; and commercial turkeys will range little and need large amounts of purchased feed.

Understand the type of turkeys you are raising, understand their needs, and adjust your management of the turkeys accordingly.

An old Italian proverb says, "*The eye of the master fattens the lamb.*" Observe your turkeys and adjust your practices, and you will be sure of great success.

5

Incubation

FERTILE EGGS REPRESENT THE SEEDS OF LIFE; they are designed to hatch. While they require certain conditions to grow properly, fertile eggs will often hatch under poor conditions. Don't be shy or concerned about your lack of experience — anyone who follows the suggestions in this chapter can have success hatching.

Eggs are pregnant. Peter Brown, a.k.a. The Chicken Doctor, of First State Veterinary Supply was the first person to tell me this. Peter's insight is profound and helps poultry people better understand all the particulars that can impact the level of success when planning to incubate fertile eggs. In order to achieve the best results, keep this in mind while planning for the collection, storage, and management of fertile eggs for hatching.

Collection of Eggs

Turkey hens seek out a private nesting site in which to lay their eggs. When collecting eggs, it is best to avoid disturbing a hen turkey on a nest as this may cause her to seek out a new, "safer" location. Interruptions also may affect a hen's laying cycle and cause her to slow or stop egg production — this is especially true near the beginning and end of her annual lay cycle. To help the turkey hen feel the nesting site is a secure location, take care not to disturb the nest, nesting materials, or the site around a nest. Hen turkeys can be very sensitive to disturbances of a site and may abandon one. When you consider that in the wild the hen will spend 28 days covering her eggs, unprotected, it is easy to understand her caution.

For best hatching results, collect eggs daily to prevent temperature fluctuations from killing embryos and ensure the eggs remain undamaged. When eggs become chilled for several hours, they often fail to produce live embryos. Traditionally, eggs in winter were collected several times during the day — to prevent freezing and heating and cooling as hens entered and left nest boxes. Embryos in eggs that warm and cool one or more times often fail to grow.

Whether you collect eggs once a day or several times, be sure that the time(s) of collection do not cause other problems. Avoid disturbing hen turkeys by timing the first collection of eggs when few hens are on the nests preparing to lay eggs; sometime between 10 A.M. and noon is best. Late in the afternoon, near dark, is the best time for the day's final collection — or only collection if collecting just once a day. Collecting at the end of the day prevents eggs left on the nest, exposed to the cool temperatures of the night.

Take care to prevent damaging the shells of the eggs during or prior to collection. When collecting eggs, place them carefully into an egg basket or egg cartons for safe transportation to the storage location. Nest boxes should be lined with plenty of cushioning material such as pine shavings or straw.

Eggs should be appraised for the relative thickness of their shells. Simply crack open one egg and note the amount of effort required. If the eggshell is too thin, the turkey hens need oyster shell or other calcium supplements free choice (always available so that the hens may consume as needed, balancing out their own nutritional requirement).

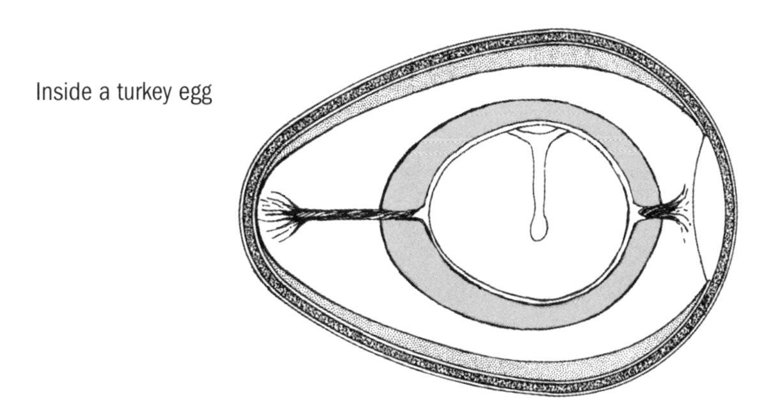
Inside a turkey egg

Storage of Fertile Eggs

From the time it is laid until it is placed into an incubator or under a setting hen, the embryo inside an egg is growing very slowly. When an egg is warmed to 99 or 100°F (37.2–37.7°C), the embryo begins growing quickly. If the embryo is not growing, it is dead. Storage of fertile eggs is about maintaining a constant temperature to help ensure the viability of hatching eggs.

Location is important. Eggs being saved for hatching should be stored in a cool location, out of direct sunlight to avoid temperature fluctuations. When sunlight shines on eggs, it tends to raise their temperature, possibly causing the embryo to begin to grow and then die as the temperature cools down in the evening. Ideal locations include a cellar, closet, an office, or a spare bedroom with window blinds drawn closed. Store eggs for hatching between 50 and 60°F (10–15.5°C).

An insulated cooler works very well to keep the eggs at a stable temperature with a constant humidity. Simply place the cooler in a cool location where it will not be disturbed. There is no need to add ice; simply use the insulated nature of the cooler to maintain a constant temperature and humidity level.

A consistent humidity level is important to help control the amount of moisture present in the eggs at the point of setting them in the incubator. Eggshells are porous; eggs will lose a little moisture before hatching. Try to avoid losing too much moisture, because the paper liner of the shell then becomes too tough and the poults will not be able to get out of their shells. If you have no way of measuring humidity, you can weigh the eggs when setting and expect them to weigh 5 percent less on the 12th day of incubation and 10 percent less on the 24th day. Storing eggs at a constant humidity level allows all the eggs to lose moisture at a similar rate, improving hatch rates.

Eggs stored for hatching should be rotated once per day. This prevents the embryo from sticking to the side of the shell, or from becoming dislodged from the yolk. Eggs can be turned by hand individually one or more times per day. When using a cooler, you can place a 2 × 4 board under one end of the cooler, switch the board once per day, and thus rotate the eggs with little effort.

Culling Eggs and Pre-Incubation

Fertile eggs, well stored, will hatch even when saved for as long as 3 to 4 weeks. For best results, save eggs for just 10 days, although saving for 2 weeks makes it easier to plan for hatches. Bob Hawes of Maine has noted that hatchability is reduced by 4 percent and hatch is delayed by half an hour per day beyond 4 days that eggs are saved. Since it takes 4 weeks for incubated eggs to hatch, when using a single incubator you can save eggs for 2 weeks, set a batch of eggs, sell your eggs for the next 2 weeks, and then save eggs for 2 more weeks for the next setting.

For optimal results, allow eggs to warm to room temperature for 4 to 6 hours before placing in the incubator. Freshly laid eggs should cool to room temperature for this amount of time as well. While setting eggs, realize that culling the flock starts with the eggs. Discard all misshapen eggs, eggs that are long and thin, eggs that are too round, eggs with rough or thin shells, and heavily soiled eggs. Egg size may be increased in a flock over time, by selecting only large eggs for setting. This does not work well, though, in mixed-age flocks as older hens tend to lay larger eggs than young hens. Be careful in selection, as hens may lay fewer total eggs per year as egg size increases and hatchability reduces when egg size becomes too large.

Be aware that a warm and moist incubator is an ideal breeding ground for bacteria. A dirty incubator, infected with bacteria, is not likely to produce hatches with any level of success. Heavily soiled eggs, or those with cracks, may introduce problems into the incubator. Eggs with cracks in their shells are vulnerable to bacteria penetrating them, causing them to rot and possibly explode in the incubator. Dr. Al Watts of Washington state once told me that he drips the wax of a candle along the cracks and is able to hatch from eggs with light cracks. I've tried this and it works. Heavily soiled eggs bring large doses of bacteria into the incubator; with their pores largely blocked, they are less likely to produce a live poult.

If using shipped eggs, rather than eggs from your own flock, you can help increase the number of poults that will hatch. The jarring eggs experience during shipping leaves the yolks and whites of the egg loose and "unformed" — crack one open onto a plate to see how the yolk and the white spread out thinly across the plate. If you let the shipped eggs sit overnight, the yolks and whites will "re-form" and gradually come to room temperature; if cracked open onto a plate, the whites will stand up a little and the yolk will stand firmly on top of the whites. Breeders for many decades have made it a practice to let shipped eggs, or eggs that were heavily jarred, rest overnight before placing into the incubator. Try this for yourself and you will see an increase in the number of poults that hatch from shipped eggs.

Incubator Location

Incubators must be set up in a location without direct sunlight — to prevent temperature fluctuations — and with a constant, but moderate, temperature. Fluctuations in room temperature often result in poor hatches because the incubator heating system is not able to warm up as fast as room temperatures may drop, and the incubator may become too hot when room temperature rises. Seldom does an incubator perform correctly in a garage or barn — nighttime temperatures in early spring are just too low for the incubator to maintain internal temperature. For some reason, incubator manufacturers never seem to have thought of insulating their products. My friend, Raymond Taylor of Virginia, built an insulated, small closet in his garage for his incubator, and this proved very successful at allowing the incubator to maintain a nice, even temperature. I usually place a blanket over my incubator, without blocking the circulation vents, and joyously have found lower electric bills and very successful hatches as a result.

The best location for an incubator is in a basement or cellar. Such a place usually has an even daytime/nighttime temperature, no direct sunlight shining on the incubator, and a good level of humidity. In fact, sometimes you will not need to add any moisture as humidity levels will be ideal for hatching. A closet can be used successfully, as can a spare room — as long as any windows are closed and shaded.

Incubator Setup

The first thing you should do after choosing a location for your incubator is clean and disinfect the incubator. Actually, cleaning and disinfecting the incubator is also the last thing you should do at the end of each year's hatching season. This way, there is no material upon which bacteria can grow prior to next year's use. I use a vacuum cleaner to remove all shells, down, and dust. When cleaning at the end of the season, run the incubator for a day without any water so all material dries prior to vacuuming. Follow up by scrubbing the trays and the inside of the incubator with a disinfectant, and then spraying with disinfectant. Scrubbing is the best method to remove dried matter.

While a new incubator may not need scrubbing, spraying the interior with a disinfectant is a good practice. Pine-Sol, Tek-Trol, or a bleach-and-water solution are some of the disinfectants I recommend. Choose one and use according to directions — when using Tek-Trol, take care you do not breathe in the fumes as they are irritating to lung tissue. Spray the incubator and then turn it on and run it until the disinfectant has dried; at that point you will notice that most of the smell of the disinfectant is gone. Do not set eggs until the disinfectant has had a chance to evaporate as the fumes can cause poor hatching results.

Turn on the incubator at least 5 days prior to setting the first clutch of eggs. During this time, adjust temperature to manufacturer specifications and check several times each day to be sure it is being maintained. Generally, still-air incubators run at 101°F (38.3°C) and forced air at 99.5°F (37.5°C). Measure and adjust humidity as well. Once everything is at the proper adjustment and holding steady, you are ready to set the eggs. If not, the 5-day period should give you enough time to make adjustments or figure out the cause of any problems.

As I've learned, a blanket placed over the incubator helps reduce heat loss. If you plan to do the same, do so during these first 5 days to ensure temperature is not affected. Certainly, a blanket can be especially useful to maintain warmth in the incubator should a power outage occur.

Always follow the incubator manufacturer's instructions for operating your incubator. Manufacturers spend a good deal of time and effort in designing their equipment, and they offer some excellent troubleshooting information.

If you are using an older incubator, check to see if an operator's manual is available on the Internet. If the incubator uses a wafer thermostat, you can retrofit a new digital thermostat for more accurate temperature control. At the very least, purchase a spare thermostat for wafer-controlled mechanisms and keep it handy near the incubator.

If you wish to produce a number of consecutive hatches, you may consider running two incubators — the first is used to incubate the eggs for the first 3 weeks, and the second is used only for the last week to hatch the eggs. The advantage of this system, using one incubator for incubation and the other for hatching, is that successive clutches may be set while all the debris of hatch is confined to one incubator. This means that you need only clean the incubator at the start of each season and the hatching incubator after each clutch has hatched. Hatching incubators should be running for 5 days and up to proper humidity and temperature levels before eggs are moved into them.

Incubator Temperature and Humidity

Incubator temperature has several effects on successful hatches. Running the incubator temperature too low will delay the hatches; running it too high makes the hatches come early. My old neighbor, Paul Seymour of Virginia, has observed that lowering the incubator temperature by half a degree results in more females and raising it half a degree results in more males. The jury is out on exactly how this works, but it seems that the lower temperature results in fewer of the male poults hatching, and just the opposite with the higher temperature. Studies show that poults left in the incubator for 24 hours after hatch suffer less from heat stress as adults.

Proper incubator humidity level is important. Turkey eggs do best when incubated at 60 percent relative humidity for the first 24 days and then 70 percent for the last 4. Eggs should lose a total of 5 percent of their weight during the first 12 days of incubation and a total of 10 percent by the day of hatch. Weighing eggs at the time of setting and again on the 12th day of incubation is a simple method of measuring moisture loss.

When humidity is too high during the entire incubation period, the poults will hatch but their navels will not seal properly, thus leaving them vulnerable to bacterial infections. High temperature will also cause a similar condition of the navel. High humidity can result in many poults that pip but then "drown" from the moisture that collects at their nostrils. The poults that do hatch will have a "pasty" appearance because not enough moisture evaporated from their down when they hatched, causing the down feathers to cling to one another and lie flat instead of being fluffy.

When humidity is too low for the entire incubation period, the poults pip and fail to hatch — the eggshells are too tough because their membranes are dry. Those poults that do hatch will have shriveled, thin legs and be more prone to dehydration the first 2 days in the brooder. The poults will have less yolk in their bellies, as more has evaporated prior to hatch, and this will make them weak if they do not learn to eat within the first day or so.

Most incubators have a tray or an attached container to which you add water to control humidity level. As the water evaporates, it leaves behind mineral residue. In some locations, poultry people may choose to use bottled or distilled water. I have not observed any differences in water used for incubation, but in all cases the water used should be clean, clear, and free of odors. City water may cause some issues as it can contain chemicals, such as chlorine, which will evaporate as the water evaporates. To keep the incubator environment clean, a friend of mine adds a teaspoon of bleach to each gallon of water; while this works, it does tend to corrode the metal parts of an incubator. I often add a teaspoon of Pine-Sol to each gallon of water — it keeps down bacteria, makes the incubator smell nice, and does not corrode metal. A caution: Adding too much bleach (or Pine-Sol) can reduce hatches; remember, the growing chicks breathe through the porous eggshell.

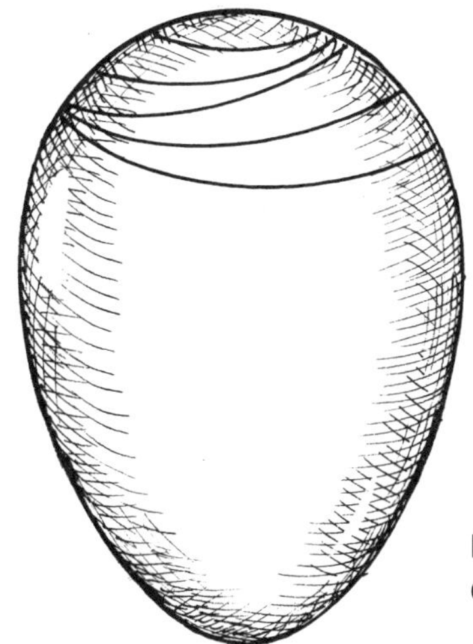

Effect of humidity loss on air cell during incubation

Egg Position and Turning

Incubation takes 28 days. During this time, eggs should be first placed pointed end down and then rotated every 8 hours. The pointed end must be down in order for the air cell to remain at the top of the egg. If the air cell develops malposition from being placed downward during storage or incubation, the poult will not be able to hatch — the relationship between the air cell location and the poult's beak at the time of hatch is critical to extrication from the shell.

Many debate whether eggs are turned when incubated by a hen. Some may notice certain hens turning their eggs, while others will notice no such effort. But you can be sure that the movements of the mother hen from the nest and her resettling have the effect of rotating the eggs. It is remarkable that some people have success when placing their eggs pointed end up — their success is not due to correct method, but rather to luck and is an excellent example of the eggs' ability to hatch under less-than-perfect conditions. Whether rotated or not, pointed end up or down, if someone has success doing the opposite of recommended ways, it only proves that nature has imbued eggs to hatch — even under less-than-ideal conditions. For best results, stick with proven methods that work; turn your eggs and place them pointed end down.

When using artificial incubation, turn eggs daily, as much as five times per day for optimum results. Essentially, turkey eggs should be turned often enough that they are never left unturned for more than 8 hours at a stretch. Turkey eggs benefit from frequent turning, as this prevents the embryo from growing in an improper position or sticking to the side of the eggshell.

Candling

As the turkey poult grows, its body takes up more room inside of the shell. Candling is the act of holding an egg up to a light and allowing the light to illuminate the inside — showing you if there is a growing, developing poult inside. By day 10, some of the nerves and blood vessels of the poult can be seen spread across the inside of the egg; when the egg is held up to a light, the poult's body will appear as a smallish blob. As the poult grows, it becomes easier to identify with candling.

There are many opinions as to when you should candle eggs. Most sources recommend candling on day 10 and again on day 24 or 25. Any nonfertile eggs or eggs that have stopped developing should be removed. Leaving nonfertile eggs can cause uneven temperatures inside the incubator and carries the potential of releasing gases; either action will decrease successful hatching of the remaining eggs. Remove cracked eggs as well; they can allow bacteria inside the shell and have the potential to explode when handled, leaving a stinking goo all over everything inside the incubator.

Play it safe, and remove infertile eggs when you find them. My own preference is to candle the eggs on day 14 and day 25 as this fits with my weekly tasks, movement from the incubator to the hatching incubator, and makes it easy to differentiate the growing eggs from eggs that have died.

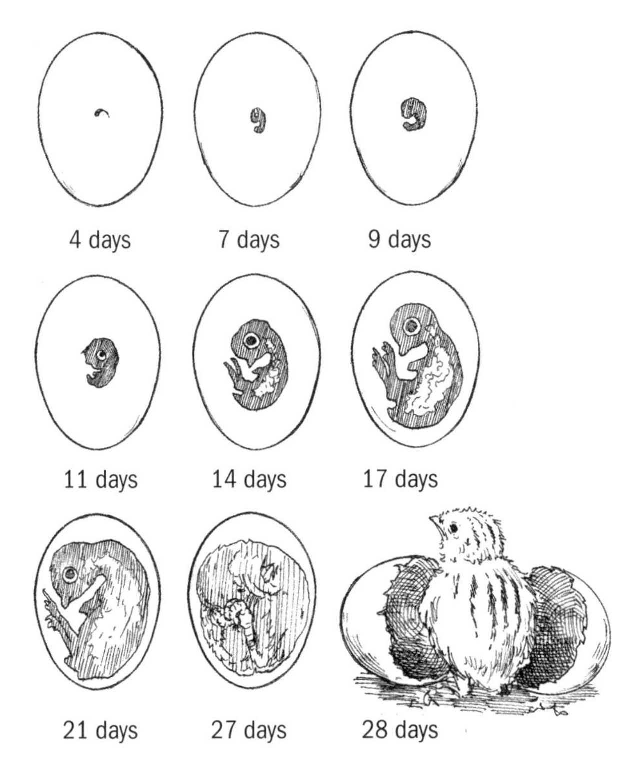

Turkey egg development. The poult grows and develops using the egg white as a source of nutrition. During the last week of growth, the poult's body grows around the yolk – giving a food source for the first few days of life.

IN CASE OF POWER OUTAGE

It is important to know that a power outage does not necessarily kill a clutch of incubating eggs. If the power goes out during incubation, cover the incubator with several blankets to hold in the heat, and by all means, do not open the incubator! Many times a power outage, even one lasting 2 or 3 days, will delay the hatch by 2 to 4 days. Simply wait for the power to come on, and a week later candle the eggs to determine if they are still growing.

When to Open the Incubator

Rotation stops on day 25. If eggs are to finish in a hatching incubator, the move time is now. If using a single incubator equipped with a lower, hatching tray, move the eggs to the bottom tray for hatching. At this time, most incubator manufacturers recommend raising the incubator humidity a little for the last 3 days prior to hatch — this helps ensure the poult can penetrate the paper membrane. From this point on, do not open the incubator and risk reducing humidity. If you have to open the incubator during these last 3 critical days, make it quick and spray a bit of mist into the incubator, with a spray bottle as you work. When the poults start hatching, do not open the incubator. Many of us wish to hold that first poult, but the sudden drop in humidity may result in many others pipping and not hatching. Resist temptation — wait for the others to hatch!

I always plan to open the incubator on day 29 instead of day 28. The poults grow and develop using the egg white as a source of nutrition. During the last week of growth, the poult's body grows around the yolk, sealing up just prior to hatch. You will notice day-old poults have a navel on their lower abdomen. It leads to the yolk that is a source of nutrition and moisture for the poult for the first 3 days of its life. Leaving the poults in the incubator for one extra day allows any stragglers a chance to hatch before the door opens and the humidity drops. This practice has the added benefit of heat tempering the young poults, making them less susceptible to heat stress as adults. Since the first-hatch poults have the yolk to live off of for the first 3 days post-hatch, they will not suffer or be negatively impacted by the practice of opening the incubator on day 29.

Poult Defects Caused by Improper Incubation

Normal position for poults and chicks prior to hatching is head to right, under wing. Any other position leads to embryos that fail to pip or pip but cannot hatch.

The most common malposition is head to left, either over or under the wing. This malposition may be heritable — not that such a poult hatches on its own, but if helped it may and then it will pass along the trait. Guard against such potential genetic defects to protect the reproductive soundness of your turkey flock. Malposition of the air cell causes most of the cases of a poult growing to full term without hatching.

The poult must grow and be in proper position or its struggles will not extricate it from the shell. In nature, such birds do not hatch. Don't help them out of the shell! First, you do not want to increase the incidence of malposition of air cell in your flock. Second, the struggle to hatch causes the poult's metabolism to speed up — in particular, the circulatory system must work to match an increased level of exertion. This is necessary to have healthy poults that grow into productive, healthy adults. Helping "weaklings" out of the shell seems kind, but works against nature and handicaps future generations.

Crooked toes are another defect that results from poor management during incubation. Crooked toes in turkey poults occur when the poult spends too much time attempting to hatch — this usually happens when humidity is low during the hatch, or up until hatch, and the membrane of the shell becomes too dry, making it more difficult for the poult to penetrate. Poults that have been unable to remove themselves from their shells unassisted often have tendon issues. Once poults exert themselves and their metabolism increases, the tendons start to set — if they do so while the poult is in the shell, then crooked toes and even bow legs can and do usually result.

Identification and Tracking

Once eggs are collected, they become very hard to tell apart. If you are hatching poults from more than one mating, you will need a method of identifying them based upon ancestry. Identification starts by marking the eggs according to mating. You can use a pencil to write on the shell, and identification can be as simple as "A," "B," and "C" or more complete such as "Bourbon Red Mating 2." Avoid using a felt-tip marker on egg shells as the fumes could penetrate the porous shells.

At the very least, keep a sheet of paper or a logbook near the incubator. Use this to log the number of eggs set for each mating in the clutch, the number of eggs for each mating that are fertile, and the number of poults hatched per mating. The log should include the date set and the date of expected hatch. Maintain a record of these details each time you operate the incubator. I make a crude drawing of each hatching tray with the number of eggs of each mating. This way, when the poults hatch I know how many to expect in each tray and which mating they are from — the marks on the eggs will be very hard to read once the eggs have

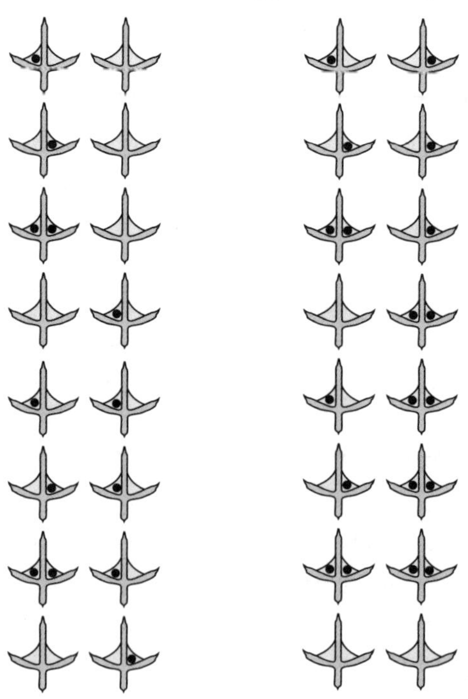

This diagram shows how to track your chicks' parentage with 16 toe-punch options, each indicating a different hen-tom pairing.

CORRAL THOSE ESCAPE ARTISTS

A few tips during poult removal are in order. Poults are often ready to run out and explore the world. Never expect them to sit idly by as you reach in to grab them. In incubators with several hatching trays, remove the poults from the bottom tray first. In this way a group below is not free to escape by jumping out of the tray to explore the world. When a tray has cages, first remove the poults that are loose on that tray. Have covers for the boxes in which you place the poults, or expect to have escapees. Make sure the boxes are ventilated — this will allow you to place a large number into one box, where body heat will keep them warm and ventilation will bring them fresh air.

hatched. A log will prove to be a surprising bit of useful information and can even help when planning hatch dates for the next season.

Once the poults hatch, mark them according to the mating from which the eggs originated. Since the poults will mix together quickly and appear nearly identical, the best method is to prevent poults from different matings from mixing together before they have been marked. This can be accomplished by hatching poults of only one mating at a time, by hatching only one mating per incubator tray and marking the poults of one tray before mixing with other poults, or by creating little wire cages to place on each tray to contain one batch per cage.

The method I prefer for marking poults is the toe punch. Toe punching is the practice of punching out a portion of the webbing between the toes of one or both feet. Since each foot has a web between the outside toe and the middle toe and a second web between the middle and the innermost toe, you have two webs per foot — or four webs in total. By punching a combination of webs, in groups of two or three punches, by punching all or no webs, and by using single-web punches you can identify up to 16 different matings. (See toe punch chart on opposite page.)

To toe punch, hold the poult cupped in your hand, right side up, his little head sticking out between your thumb and first finger, and use the other hand to spread the toes and operate the toe punch tool (available from most poultry supply companies). You want to make a clean hole in the webbing, not too close to the juncture of the two toes (as some bleeding can occur if the veins at the juncture are cut). The hole must be clean, and the bit of skin punched from the webbing must be removed or the mark may fail and the hole grow over and seal up — rendering the process useless. I often punch two holes adjoining to make one larger hole to prevent loss of the mark. You may also punch clean through the front edge of the webbing if you like. This leaves no hole to catch on sticks, preventing future injury as adults, and leaves a marked web appearing less deep than other webs.

Toe punching should be done as the poults are removed from the incubator. Have two or more boxes ready in which to place the poults as you remove them from the incubator. Place all the poults from one mating into one box and close the incubator. Count the poults in the box and make note on your log. Toe-punch this group of poults, placing each one into the other box once it has been marked. As each group is marked, open the incubator and remove the next group, placing them

TURKEY IDIOMS

Talking turkey. To discuss a problem in a serious way with the intention of solving it; to talk frankly.

Cold turkey. The act of doing something immediately, without tapering or cutting down gradually.

Turkey shoot. In war, a fight that is sure to be a one-sided victory.

Like turkeys voting for Christmas. When voters support ideas that are certainly not in their best interest.

Turkey(s). A form of endearment that playfully insults; a group of people that joke around.

Strut. To show your skill in something that involves movement, like dancing.

Gobble. To eat something completely and rapidly.

Turkey trot. A dance inspired by the movements of turkeys. Introduced about 1909, its chief popularity was in being denounced by the Vatican.

into an empty box and moving them to another box as they are identified. Continue until all poults are removed and marked.

Removing Poults from the Incubator

When removing poults from the incubator, open the incubator as little as possible to prevent loss of too much heat and humidity — this is especially important if there are still poults yet to be removed. Have a trash can or bag nearby and remove the debris of hatch, such as eggshells and any poult that may have hatched and then perished. I usually give any unhatched eggs or eggs that have pipped, with live poults in them, 24 more hours to come out of the shell.

Sometimes a poult will pip and never hatch. I follow nature's example in that I never help such poults out of the shell. I have given many good reasons previously, but let me now share a personal experience so you understand my position. My mentor in poultry has a line of bantam Leghorn chickens — in fact he is one of the originators of a particular color for these bantam Leghorns. He has had the line for 60 years and at some point, he started helping chicks out of the shell that did not make it out on their own. While I was breeding his line, I was troubled by the high pip rate and low hatch rate. I made a hard decision and stopped helping the chicks out of the shell. The result was 3 years of letting half or more of the hatch perish in their shells — but the end result was that the line started hatching on its own again! My mentor's kindness had spread a heritable defect across his line; by following nature's example, I was able to breed the fault back out of the line. Save yourself the trouble; from the start do not help fix a genetic fault into your line of turkeys through a misapplication of kindness.

When a poult is stuck in its shell, I usually opt to euthanize it (after giving the 24 hours I mentioned above). Many people have shared their methods, but the method I follow is that of drowning in warm water. I place the eggs with the pipped poults in a glass jar and then fill with tap water warm to the touch. After a few minutes the poults have perished. If a poult is injured and not likely to survive, I will sometimes put it down as well; I use the same method but modify it by holding the poult's head under the water using a stick. There may be a better way to euthanize a poult, but I have yet to find it.

◆ ◆ ◆

Important points to keep in mind when incubating are:

- Turkey eggs for hatching should be gathered at least once per day and stored at 50 to 60°F (10–15.5°C) in a cool place with a relatively constant temperature. Turn stored eggs once per day.
- It is best to set the eggs within the first 10 to 14 days after collecting, though a good percentage will hatch, if stored properly, even when they are 3 to 4 weeks old.
- If you have no way of measuring humidity, you can weigh the eggs when setting and expect them to weigh 5 percent less on day 12 and 10 percent less on day 24 after incubation begins.
- Do not open the incubator after day 25, to prevent a drop in humidity that can result in fewer successful hatches.
- Turkey eggs should be turned at least once every 8 hours.
- Poults that cannot make it out of the shell on their own may carry hereditary defects, and so should never be used as breeders — it is in the flock's long-range best interest to euthanize such poults.

As you can see, nature should guide your efforts as closely as possible. During incubation and hatching, the best methods are to do as the hen turkey would. You should not help poults of their shells, you should dry them using the warm moist heat of the incubator, and you should turn the eggs for best results.

6

Brooding

IN ORDER TO HAVE ANY SUCCESS in a turkey enterprise, it is necessary to keep in mind the amount of time it takes to grow a turkey to reach a desirable state of development so that the turkey can fulfill expectations. It takes 26 to 28 weeks to mature a heritage turkey from hatch to proper finish — turkey toms will grow to finished size in 28 weeks and young heritage hens will grow to a finished size in about 26 weeks. It takes 16 to 18 weeks to ready industrial-type, broad-breasted turkeys for processing. The American Poultry Association Standard of Perfection suggests mature weights for young toms and hens at thirty-four to thirty-six weeks of age — important for those who plan to show or exhibit their turkeys.

When raising turkeys to produce meat, secure the stock from a source known to produce good specimens for the table, with a good measure of consistency in the young birds. For farmers whose meat customers demand a large, broad breast and the appearance of an extremely plump carcass, the commercial broad-breasted turkeys will likely be the only choice. Keep in mind these birds have been selected to perform in the controlled environment of a large building; they will not do well on pasture. When your customer base demands flavor of the highest caliber, choose a heritage turkey variety that has a reputation for distinctive good flavor, such as Bourbon Red, Black, Bronze, Midget White, or Royal Palm — these have all won flavor comparison tastings.

For those who wish to exhibit, purchase turkey poults, eggs, or adult stock from flocks bred for quality appearance. When choosing turkeys for pets, or for use in a diversified small farm production system, heritage turkeys are recommended for their robust immune systems.

Planning is the surest road to success in any turkey enterprise.

Imitating Nature

Brooding turkey poults is the act of taking infants and providing them the conditions necessary to grow to be happy, healthy, and well-balanced adolescents. In beginning this process, you take on the role of parent and must provide them with food, drink, shelter, and warmth.

PLAN FOR SUCCESS

All turkey entrepreneurs should have a clear idea of the system of production they plan to use, the purpose of the enterprise, and the products they will produce before starting the enterprise. What product(s) do you hope to produce from your turkeys — eggs, dressed whole birds, turkey bacon, colored feathers, breeding stock? Where will you market these products? What are the expectations of your customer base? How will you raise your turkeys? All of these factors should help define the right choice of stock, bred to perform and produce certain products within a particular system of production. Your choice of stock and source can mean the difference between success and failure.

To illustrate the point, which is better — a Chevy Corvette or a Ford F-250 4×4 truck? Which better serves your purpose? Well, the 'Vette is faster (on paved roads), handles curves at high speeds, and looks very cool. The truck handles winter driving better, can haul loads of feed and equipment, and can be driven across the fields with little fear of getting stuck. If you know you need your vehicle to haul lots of heavy stuff across winter fields, a Corvette is the wrong tool for the job. Likewise, you should take care in deciding which source of stock will provide you with turkeys designed for the system of production and the end product that you plan to sell.

All efforts are, and should be, an imitation of natural brooding by the hen turkey. A hen turkey provides warmth, shelters her poults, teaches them what to eat, and shows them where to drink. The hen talks to her brood, using the tone of her voice to comfort the poults, alert them to danger, to call them to feed, or to call them to her to reassemble the group. The hen's relationship with her brood actually lowers the stress levels of the poults while providing them with all that young turkeys need.

There is much more to the nature of a brood of turkeys and their mother. A hen turkey is ever present to warm poults as they need it — note that the poults are not constantly being warmed, but may explore the world and return frequently to warm up. The hen must shelter her brood from rain as the combination of moisture and chill can be lethal to the poults as long as they are still in their down. A natural-sized brood of poults ranges from 6 to 16. Poults tend to stay together instinctively. Should one wander out of sight of its fellows, it will immediately give a call to find and rejoin the group.

Natural Brooding

For those readers with a small rafter of turkeys, I recommend letting a hen turkey raise a brood of poults for you. A hen turkey will lay a clutch of eggs all in one location. If the eggs remain undisturbed, there is a good chance the hen will go broody and will soon sit on the nest and cover the eggs — spending the nights on her eggs and remaining for the next 28 days with only short, half-hour or hour-long breaks to eat and drink or stretch her legs.

Warm conditions and the sight of a full nest of eggs will induce broodiness in many turkey hens. It is also possible to induce broodiness. A friend's mother used to place a turkey hen in a clothes basket next to the kitchen stove with another clothes basket on top to keep the hen inside. She would insert a clutch of goose eggs, and within a few days the turkey hen would go broody. The family raised geese for profit, and turkey hens were invaluable for the ease of inducing broodiness and for brooding the goslings after they hatched.

A hen that has recently gone broody on her own should be moved to a secluded area. This will give her a quiet location where she can remain undisturbed. It will also prevent other hens from climbing into the nest with her, possibly driving her out or breaking eggs — or adding eggs to the clutch which, because they are days of incubation behind the original eggs, have little chance of hatching.

The best time to move a hen is at night. By moving the hen at night, in the dark, she will be less stressed by the experience and more likely to accept the new nest site. Enclose the new site, containing the hen so that she must remain. It is also a good idea to give her a few eggs, or fake eggs, so she can sit on them. Wait a few nights before switching from the fake eggs to the ones you wish to hatch. Remember, if you must disturb a broody hen, always do so at night.

The Nest

The nest location should be clean and dry as well as secluded. It should be enclosed so that day-old poults cannot escape, and should have no holes into which poults could fall. The hen and the poults should have food and water available, placed low enough to be accessible to the poults. The brood should remain in this location for at least a few days after hatching, giving the poults a chance to learn to eat and drink and grow strong before being introduced to the rest of the flock. In the wild, the hen turkey naturally seeks out a secluded location to lay and incubate her eggs, returning to the flock after the poults are a few days old.

Turkey hens make great broodies and better-than-fair mothers. It is enjoyable to watch a hen turkey and her brood and I highly encourage natural brooding to anyone with a small group of turkeys. For the producer who needs to incubate as many eggs as possible, however, it is better to use artificial means to hatch and rear young turkeys.

Before Brooding

When hatching your own poults in an incubator, do not remove poults until they are dry. Evaporation of moisture has a chilling effect on a wet poult or chick — just as sweating cools down human bodies. The warm, moist air in the incubator allows the moisture on the down of the poults to wick away without robbing the poults of body heat.

I prefer to remove poults a day late, as the extended period of time gives stragglers a chance to hatch. In nature, the entire brood hatches within a 12- to 48-hour period so that they may all leave the nest together. The extended stay in the incubator also has the advantage of helping to condition the birds to heat as adults — resulting in fewer cases of heat stroke during summer months.

Once poults are ready to leave the incubator, prepare boxes to organize them and move them to brooding facilities. If you plan to hatch poults from multiple matings and mark them according to their pedigrees, now is the time to toe-punch the poults (see Identification and Tracking in chapter 5).

If you plan to ship or transport poults for any significant time (such as half an hour), use cardboard or plastic shipping boxes. Such boxes should have 10 × 12-inch (25 × 30 cm) compartments that are 8 inches (20 cm) high. It is ideal to place only 15 to 18 poults per compartment, as 15 is the minimum number needed to maintain body warmth and 18 is the maximum number per compartment to prevent losing any poults to trampling.

The Brooder and Brooder House

The building for housing poults with brooders should be clean and ready, up to temperature and running for a few days to ensure everything is working properly. A brooder house of 10 × 12 feet (3 × 3.6 m) is ideal. Such a building will brood up to 150 poults from day-old until they reach 8 weeks of age. For optimal results, 150 poults seems to be the maximum number of any one group. Larger groups tend to suffer more casualties due to pileups and overcrowding issues such as feather pecking and cannibalism.

Build the house on skids to allow easy movement so that each brood may be raised on clean pasture. The building should have 1 square foot (0.09 sq m) of window space for every 10 square feet (0.9 sq m) of floor space to permit ample ventilation; it should have a sound roof and be free of drafts. Allow a minimum of 4 inches (10 cm) of roosting space for each poult.

Locate a brooder more or less centrally, and round the corners with boards or a **brooder ring** — this removes corners in which poults can

BROODER TEMPERATURE

When brooder temperature is ideal, the poults will spread out with some under the heat source and some outside it. When too hot, the poults will tend to lie in a ring as far from the brooder as possible (or, when the heat source is located near one end of the pen, they will lie at the other end). When the brooder temperature is too low, the poults will huddle in a tight group underneath.

Given the chance, after the first week, the poults will regulate their own body temperature.

WATCH TEMPERATURE FOR THE FIRST WEEK

Too cold. The poults bunch under the heat source.

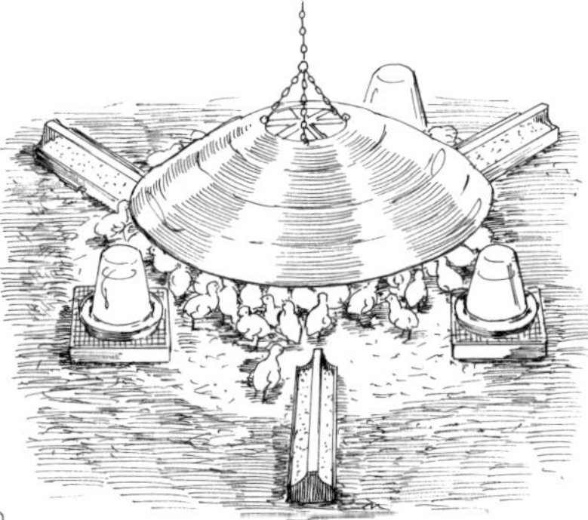

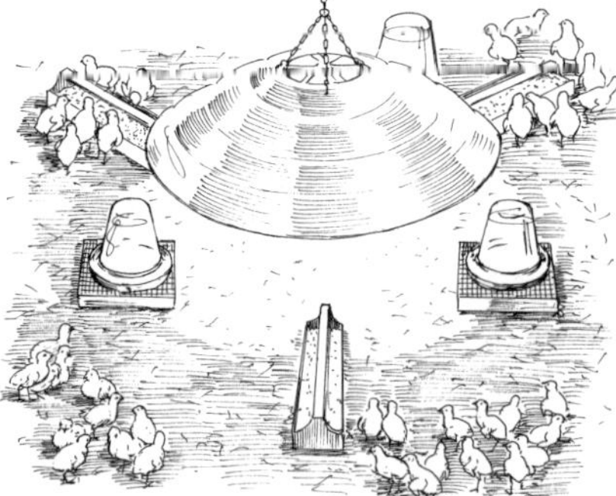

Too hot. The poults have retreated to the outer edge of the ring.

Just right. The poults are evenly spread out around food, water, and the heat source.

become trapped by other poults. Corners are the site of most brooder pileups. Use a round barrier of cardboard or wire, shaped into a ring around the brooder, to keep the poults near the brooder, feed, and water for the first week. It can be moved back gradually as the poults grow — enlarge the ring each week to give the poults more room. Encourage them to spend more time outside the brooder by moving the waters and feeders farther out as well.

Proper Use of Heat

There are various types of brooder heaters available — heat lamps, lightbulbs, electric, propane, and kerosene heaters. Whatever you use, take care to prevent fire. Suspend the heat source for the brooder about 18 to 24 inches (46–61 cm) above the bedding material. Do not overheat the bedding, as it can catch fire. Brooder temperature should be 95°F (35°C), and the poults should have the ability to move away from the heat source to prevent overheating. The general rule of thumb is to reduce the temperature of the brooder by 5°F (3°C) per week — though many producers maintain one brooder temperature and simply give the poults more space away from the brooder.

The brooder should contain more than one waterer and feeder. Spread these out evenly so that all poults will have equal access to food and water. Place waterers on platforms of screen or slats nailed to a wooden frame to raise the waterer above the shavings. Waterers should have a narrow lip containing the water; this style of lip prevents the poults from becoming wet and chilled, which can be fatal. Do not use open water containers — poults can drown in open water bowls and pans. Place newspaper around the feeders to prevent the poults from eating shavings.

Spraddle Leg

Caution: Never use slick paper, such as is found with some newspaper ads and magazines, as it can cause joint injuries for the young poults. The most common joint injury is a condition called "spraddle leg." Spraddle leg is a result of attempting to stand on slick surfaces; as a leg slips, the tendons on the inside of the leg can become permanently stretched, resulting in the leg turning outward. This injury happens very suddenly; the poult will never stand afterward and most likely must be destroyed.

In her book *Day Range Poultry* (see resources), Pat Foreman suggests a possible cure for spraddle leg. You need a small box, such as one for tea. On one of the long sides, about two-thirds the way down, cut two small holes in which to slip the poult's legs. On the opposite side, cut away part of the box so that the poult can poke its head out to eat and drink. Slip the poult's legs into the two holes up to its hock joint (knee), gently bend its legs as if it they were in a normal sitting position, and use a piece of medical tape to tape its feet to the box. Add a second piece of tape across its legs closer to the holes, and a third perpendicular to the first two, running between its legs. Keep the poult in this position for 24 hours and many times the poult will heal.

Primary Brooding Challenges

For the first few days of a poult's life, you need to worry about dehydration, chilling, crop impaction, wet litter, and pileups. These are the issues that most often lead to complications and death.

Dehydration

The number one killer of poults in the first 3 days is dehydration. The best prevention is to simply dip each poult's bill in the water as you place it into the brooder area. In this way, a large number of the poults will learn to drink and will teach most of the rest. Be sure to place several water fountains around the brooder so that the poults may find them easily and there is no competition to get a drink. Check the poults every few hours for the first 3 days. Any that seem sluggish should have their bills dipped in the water to make sure they are drinking.

Chilling

A good brooder, set at the proper temperature, will help prevent chilling. Check the poults every few hours for the first 3 days to ensure that they all spend time under the brooder. If a few are resting away from it, move them under the brooder by hand. Use a ring around the brooder for the first week to keep the poults from wandering too far from the heat source.

STRESS FACTORS

by Frank R. Reese, Jr.

Many challenges await the new chicks. Most can be prevented by good management and careful observation.

1. Chilling, overheating, and taking too long during transit of poults
2. Delay in getting poults on feed and water
3. Unsanitary surroundings — such as previously used turkey litter or presence of rodent droppings
4. Chilling and overheating during early brooding stages
5. Piling of poults
6. Cold or wet litter and floors
7. Placing more than one variety of poults together
8. Faulty ventilation during the early stage of brooding
9. Starvation (due to improper distribution of heat, feed, and watering equipment, and litter eating)
10. Improper lighting of the poults' barn (poults will not eat or drink in darkness)
11. Debeaking and declawing of poults (often done at the hatchery)
12. Dehydration (due to abrupt change of watering equipment or improper use of medication)
13. Crowding (too little floor space and not enough feeder and waterer space)
14. Handling or moving the flock for any reason
15. Exposure in the brooder house and on range
16. Various low-grade infections
17. Having multiple ages of poults in the same brooder, barn, or pasture
18. Eating spilled feed in litter
19. Competition between sexes
20. Extremes in temperatures
21. Sudden loud noises (planes flying over; guns going off; strangers coming into their space)

Impaction

Crop impaction is a result of eating the bedding, usually comprised of pine shavings, instead of the feed. To help prevent poults from mistaking shavings for feed, it is a good idea to place newspaper under the feeders. Some producers use boards or cardboard under the feeders for the first week or two. The idea is to prevent feed from spilling and mixing with the shavings, both so that the poults do not accidentally eat shavings and do not mistakenly view shavings as food. Newspaper, cardboard, or boards can be removed after the first week.

Bedding

The care of the bedding material is probably the single most important aspect of good husbandry during the brooding phase of a poult's life. Wet litter in a brooder is simply a "no-no." The brooder area is warm and the addition of moisture leads to perfect conditions for the breeding of diseases like coccidiosis: an organism that will attack the intestines of the poults, causing bloody stools, weakness, and then death. Moisture and warmth also cause the bedding and accumulated droppings to release ammonia. Wet bedding can cause poults to chill from their breasts even while their backs are warmed by the heat source.

The best litter is 2 to 3 inches (5–7.6 cm) of clean sand covering the floor. Shavings, straw, shredded cane, peat, sawdust, chopped hay, and ground corncobs are all satisfactory litter. If you use hay, be aware that it must be dry from the start and kept dry, as it will mold and break down quickly. Hay's carbon nature will help trap gases and odors. Stir bedding after 3 days, or turn daily; add a fresh handful to cover heavily manured areas. This simple practice helps mix the carbon material of the bedding with the ammonia-rich manure. Stirring the litter helps eliminate the poults' exposure to their own feces as well. If more poultrymen stirred the brooder litter, there would be fewer instances of brooder-related diseases.

Learning to Eat

Aside from dehydration, the greatest risk to a poult is starvation. If poults fail to learn to eat, they can die in those first few days. Use a good turkey

Shiny glass marbles will draw your poults to the feeder and waterer.

starter mash and give them all they will eat. You can dip each turkey's bill in the water and then the mash as you place the poults in the brooder. This will help many begin to peck at the mash and taste it.

Poults simply do not learn to eat as readily as chicks do. At first, place feed in many low, easy-to-access feed containers, such as shoebox lids, egg cartons, paper plates, or even pie pans. Since poults are very curious and are attracted to bright, shiny things, you can play off this natural tendency to help them find and identify their feed. Add brightly colored feed sprinkled over the top of the mash, such as oats or cracked corn for the first 3 days (using chick-sized grains).

Set colorful glass marbles in both the water and on top of the feed to attract the young birds' attention. Marbles are an excellent tool to use with young poults or chicks. Their shiny appearance will attract the curious poults and cause the natural behavior of pecking, and thus eating and drinking. Remove the marbles by the time the poults are about four or five weeks old. To disinfect marbles between batches of poults, place them inside a net bag (such as an onion bag) and then into a dishwasher.

Poults need a great deal of light too — keep a 100-watt bulb above the brooder for the first week to help the poults find food and water. Light stimulates the thymus gland, causing hormone production, contributing to both good health and proper growth in young birds. Part of the reason turkeys reproduce in the springtime is due to light stimulation: first for the parent stock, inducing reproduction, and then to the young stock that are timed to grow in sync with the increasing amount of light. Light can be an aid in preventing two other brooder-related challenges: pileups and feather pecking.

Poults should be encouraged to roost at an early age. From the first week on, most producers provide access to a low roost so that poults can fly up to explore and satisfy their natural instinct to roost. A roost also has the benefit of reducing the feeling of crowding by reducing the number of birds actually on the floor at any given time. It is important to keep the roost low — maybe 6 inches (15 cm) from the floor — to start and as the poults grow in their wing feathers, raise the roost to a foot (30 cm) or more.

Preventing Pileups

Pileups are the result of the poults' natural instinct to huddle together to stay warm or seek security. Pileups can happen at nearly any age, so it is wise for a producer to keep this in mind when raising large numbers of turkeys (when a small group of just a few turkeys piles up, serious consequences seldom ensue). Pileups occur during brooding when the poults are too cold, the brooder is too hot, and poults cannot get as far away from the heat as they would like, or when they are frightened by strange people, dogs, predators, loud noises, or brightly colored clothing.

As mentioned earlier, use boards or cardboard to "round out" corners in the brooder/brooder house for pileup prevention. Take care that you do not create spaces that can actually trap a few poults. Rounding out the corners limits the locations where a turkey poult can become trapped. Light is helpful as a preventive, as poults tend to pile up more frequently in the dark. Keep a low-wattage bulb running at night, or a red or yellow bulb, that will give off enough light to prevent pileups.

Feather Pecking

Poultry are curious creatures, so it should come as no surprise that the combination of boredom, crowded conditions, and their natural instinct to peck at new objects leads to feather pecking. The habit often ruins the tails of young poults, and if it persists, the feather sockets can become permanently damaged. In severe cases, feather pecking escalates into poultry driving horrible wounds into each other; sometimes a poult is pecked to death and cannibalized. Poults will even peck out the eyes of their penmates, leaving some birds permanently blind in one or both eyes. Pecking of eyes can extend into the adolescent stage as well.

Feather pecking is costly as well as disruptive and demoralizing. When feather follicles are injured, the ink they contain spreads under the skin. These areas then have a "blue" appearance and the birds sell poorly as dressed turkeys. It is best to manage the poults so that this does not start.

Cause and Prevention

Feather pecking is usually a result of boredom, plain and simple. It amazes me how few experienced people ever think to address boredom for young or old caged poultry. Simply hanging a few bones or sticks on strings, just high enough for the birds to reach, can satisfy their natural pecking instinct. Placing clumps of grass or weeds into the brooder, roots and all, is also an excellent practice and encourages greater foraging once the birds are on pasture. Bits of melon rind or cabbage are useful, too, but keep these objects away from the heat source to prevent wilting or rot (and remove the leftovers daily).

While boredom and crowded conditions can lead to feather pecking, manipulation of light can prevent or even resolve this problem. Poultry will not peck at each other, even when overcrowded, when their sole source of light is a yellow bug-lightbulb. I am not sure if it is the way

FEATHER PECKING

The habit sometimes begins when the birds have nothing to clean their beaks on, except the backs and tails of their neighbors. Turkeys do not like to have mash stuck to their beaks. Feeding oats, giving more space, and adding wires across the middle of feeders can help.

colors are seen under yellow light, or if something turns off in their brains. I have had awkward cases where I could not get the birds onto pasture as early as desired. By using yellow bulbs as the only source of light, I had zero feather pecking and no cannibalization. You can spray-paint regular lightbulbs yellow; however, do not attempt to do the same with heat lamp bulbs as you will cause a fire.

Dealing with the Problem

Treat wounded poults and dress the wounds so that blood does not attract more pecking. I have found commercial antipeck lotions ineffective — though other producers have success with them. An old farmer friend once suggested I try axle grease. Applied liberally to the wounded area, axle grease is a pretty effective preventative. Once birds get a taste of the axle grease, they seldom take a second taste! My favorite product for dressing wounds is Blue-Cote, a spray-on dressing. Blue-Cote's deep blue color hides the red of damaged flesh and blood, which can incite bullying, while covering the wound with an antiseptic coating.

Another issue related to feather pecking is nutrition. Feed low in protein or amino acids can actually cause feather pecking and cannibalism — the quality of the poult's feed is not meeting the needs of his body. The solution, of course, is to use a quality feed with at least 20 percent protein and preferably with some form of animal protein. Beef liver, fed raw to the poults, can also supply the missing nutrition. Give the poults a little the first day, and then more the second. Offer the liver for only 15 minutes, removing any left over so that no poult accidentally eats a spoiled piece. I feed raw meats on a piece of brown paper, throwing the paper away with any leftovers. This prevents the need to clean a feeder each day.

Pasted Butts

During the first week of a poult's life there is one more challenge to its health — pasted butt. The condition occurs when manure sticks to the down around a poult's vent. In most cases, the manure sticks to form a plug, preventing the poult from eliminating and resulting in death after a few days. Pasted butt can be caused by rich foods or can result from sticky down feathers when humidity at hatch is too low.

The solution is first to remove the obstruction and reduce the rich food — usually dairy or egg — and feed only fresh mash for a few days. Sometimes you can pluck off the obstruction, but most of the time you will need to soften with water first; do not dampen the entire poult, only the affected area around the vent. Q-tips or tissue work well to control how much of the area gets wet. Detection of pasted butt requires observation. No other practice is of more importance than observing your turkeys at any stage of their lives.

Receiving Shipped Poults

Everything mentioned so far applies equally to shipped poults or those hatched at home. There are a few other considerations, however, when receiving shipped poults. Poults can be shipped safely during the first 3 days of their lives. This is of course due to the fact that poultry hatch with a belly full of yolk, providing them with energy as well as moisture for up to 3 days. Shipping is stressful, though, and you should prepare for the poults' arrival to prevent further stress.

As mentioned in chapter 2, it is a good idea to notify your post office a few days before the poults' expected arrival date. Leave your phone number and ask for a call as soon as the poults arrive, so you can come right out and receive them. Have the brooder running and up to temperature. Have your feeders and waterers ready the morning of delivery.

Dehydration is a big danger to shipped poults for the first 2 days. Check the poults frequently and dip bills into water and then feed two or three times a day for the first 3 days after they arrive. You can administer additives to aid the shipped poults. Many feed stores sell a vitamin/electrolyte supplement that can be added to water for day-old chicks and poults. Electrolytes help reduce the effects of stress.

You should encourage hydration and ensure adequate blood-sugar levels so poults will be active and learn to eat as well as drink — they are, after all, low on yolk by the time they arrive. I like to add either honey or sugar to the drinking water for the first few days. This gives them a little extra energy, like hyperactive children on candy, so that they will more actively learn how to eat and drink and simply explore their new surroundings. To add sugar or honey to the water, heat a small portion of water and add the sweetener, stirring until it is dissolved. Put the sugary water in the waterer and add fresh, cool water. Be sure the water is only lukewarm by the time you place it in the brooder. Never serve water too hot for the poults to drink.

Maturing Poults

Poults become strong and large quickly by the end of the first 3 weeks. Be prepared to change to large, harder-to-knock-over feeders and waters. Anticipate this need and the poults' increasing quantities of food and water. Add low roosts by no later than the third week.

To give the young poults more space and fresh air, many producers use sun porches in conjunction with their brooder house. A sun porch for turkeys is any outdoor area enclosed with wire on the sides and roofing, and with a wire floor to raise the turkeys off the ground. Since turkeys take time to develop their immune systems, strive to keep them off the ground for the first 8 weeks. A 14 × 24-foot (4 × 7 m) house with a slightly larger sun porch is perfect for raising two batches of 150 poults from start to finish. You also can use a sun porch to raise broilers in the winter (see chapter 3).

Keep turkeys on sun porches for the first 2 to 2½ months. Use 1-inch (2.5 cm) wooden slats spaced 1 inch (2.5 cm) apart for a floor. This seems to be ideal to prevent leg and toe problems, even better than wire. If keeping older or mature turkeys, spacing 1½-inch (3.8 cm) slats 1½ inches (3.8 cm) apart is better (though too large for young poults). Sun porches should be larger than the house to allow turkeys a minimum of 5 square

MANAGING MATURING POULTS

Use age-appropriate practices in the sun porch to prevent problems before they occur. Use wire platforms for feeders and water founts. Keep the corners of the house closed off to prevent pileups. Stir the litter frequently each day to prevent caking and mold. Reduce brooder temperature quickly according to poults' "spread," and remove the brooder as soon as weather permits. Add larger, more secure roosts as soon as all poults are roosting. One role of the roosts is to increase the feeling that there are fewer turkeys, thus the effect of more space per bird — preventing overcrowding problems.

feet (0.46 sq m) per turkey after 2½ months to help prevent overcrowding. Two square feet per turkey is sufficient for the first 12 weeks.

After 8 weeks, feed the poults a ration lower in protein. Offer grains along with the same higher-protein mash, but in different hoppers so as to prevent the poults from billing out mash to get to the grain. Oats and corn are excellent grains for turkeys, particularly oats — good for helping turkeys grow strong bones to prevent bone deformities and feather pecking. From 8 to 12 weeks, turkeys will eat little grain, but as they grow they will consume more; as cool weather arrives, they will eat largely grain. Yellow corn is the best fatting feed for turkeys.

Observing Poult Behavior

Imprinting is the natural process of a creature associating itself with other creatures, especially parents, or to a herd, flock, group, rafter, clan, and so on. Turkey poults imprint at hatch. There is much to suggest that the vocalizations of the hen turkey and of poults before hatching forms a basis for association as well. It certainly must comfort the poult to join the creatures with which it has been communicating, upon hatching. Eye contact is very significant to turkeys, and imprinting for a familial unit begins at hatch.

For the first week, poults sleep for hours at a time, followed by short periods of brisk activity, and then more sleep. Happy, healthy poults will be active when awake and will tend to move a short distance away from your hand as it enters the brooder. Young toms are usually bolder, and will be the first to stand their ground or to peck at some new object. Lethargic poults are not healthy, and usually are suffering from something such as too little to drink, too little to eat, a bacterial infection (in the case of a poult whose navel does not seal prior to hatch), or too little warmth; it could also be ill from a brooder disease such as coccidiosis.

When a turkey poult is approached by another poult, it will drop its wings, raise its tail, and strut. Strutting, intense eye contact, lying prostrate with neck outstretched, and moving briskly away express dominance or submission — mock mounting will also occur. Dominance roles seem to be entirely interchangeable. Dust bathing will start very early — even at just a week of age. Dust bathing is usually a communal act, so do not be surprised to see several poults digging little depressions in the bedding of the brooder. Given the chance, poults will sunbathe for hours. Even adult turkeys like to sunbathe.

Turkey poults aggressively pursue and capture insects. They shake large insects to death and then eat them. A poult will try to keep its captured prize while other turkeys give chase. They eat various flies, gnats, ants, leafhoppers, black field crickets, grasshoppers, slugs, small spiders and centipedes; large worms are less interesting. Poults also eat a variety of seeds and even sand or grit.

By three weeks, juvenile flight feathers fully emerge. At this time, you will know that you have passed one of the major points in a young turkey's life. The next period of concern is when they reach eight weeks and are first introduced to pasture.

◆ ◆ ◆

All in all, the first 3 weeks of a poult's life are the most challenging. Start with good-quality stock from a source tested to be free from disease. Feed your young poults a good quality, high-protein feed (20 percent, more or less). If funds are tight, purchase a less-expensive feed after these first 3 weeks are over — don't skimp on quality feed early on as this time period lays the foundation for producing a quality turkey. Empty and refill the waterers at least once per day, preferably twice. Stir the litter in the brooder at least once per day. Each time you visit, add a handful of litter to areas that receive heavy fouling, such as under the heat source.

For those experienced with brooding day-old chicks, the primary differences in brooding poults are: poults are slower to learn where to eat; poults are more apt to pile up in corners; poults are more easily frightened; and poults are more easily chilled. Also, chicks can be used as teachers for poults. Chicks from a disease-free source, which have not yet been exposed to soil, cannot infect your turkeys with blackhead disease. They can help curious young poults learn what is good to eat and where to drink. If you are raising multiple batches of poults, you may add a poult that is 2 or 3 weeks older than your day-olds to act as teacher poult.

The day-old poults will see such teachers as surrogate mothers and will tend to follow their lead.

Success with turkeys depends on mastering good brooder husbandry. Observation is the key element to good husbandry. Watch your turkeys often, and spend time with them. Learn their natural behaviors. Identify odd behavior — such as a lone turkey standing by itself, possibly in a corner. And use the information in this chapter to help you discover the source of the malady.

You may notice this chapter is full of warnings of what can go wrong while brooding turkeys. If that makes you feel brooding can be difficult, then it has succeeded in teaching you how to address the challenges you may face and instill a sense of importance to brooder care. If you can raise your turkeys past the brooder stage up to the period of moving them to pasture, you have passed most of the biggest challenges to raising your turkeys.

The old-timers used to say, "Well hatched is half raised." To this I would add, "Properly brooded is well raised."

7

Pastured Production

PASTURING TURKEYS GIVES THE BIRDS optimal healthful conditions and the ability to express their natural tendencies and behavior. As a system of production it has a positive impact on the land, the farm's economy, and the customers. As a practice it is framed by stocking density, size of available land, and duration.

Turkeys raised on pasture are in their natural environment. They benefit by being able to express natural behaviors, exercise, and grow in a relatively clean environment where sunlight kills pathogens, clean pasture limits contact with disease pathways, and where immune systems can be naturally stimulated. On pasture, ammonia buildup is removed and the damage it can cause to lung tissues is prevented; dust levels in the air are reduced to minuscule amounts; air is fresh and moving; and natural food sources, like grasshoppers and forage, are available to reduce the feed bill while providing the freshest possible sources of nutrition.

Overview of the Three Systems

We can choose among three production systems to raise our turkeys: confinement based, pasture based, and fixed buildings with outdoor access.

Confinement

Confinement uses overpopulation, a bedding source, and a roof to help control climate and prevent the bedding from becoming too wet.

This system can work and is preferable when you are using turkeys selected for extreme rates of growth, which need very specific conditions to perform. Placing this type of turkey outside of the system for which it is developed is a little like using a Corvette to drive off-road. A Corvette can drive just fine across a flat field, but it cannot reach its top speed without a paved road, and it cannot handle large hills or drive across small logs.

Pasture

Pasture-based production relies on frequent rotations to keep the turkeys from being exposed to large quantities of their own manure. Fresh pasture is clean and full of live food for the turkeys. Typically, we would use a dense population of turkeys in this system, combined with frequent rotation.

This can result in stress. Pasture-based production relies on turkeys that can handle stress better, and which have better immune systems

PASTURING PROS AND CONS

When practiced properly, pasturing turkeys helps eliminate many of the challenges of production, such as:

- Overcrowding
- Boredom
- Lack of exercise
- Exposure to fecal matter buildup
- Exposure to dust and ammonia

It also addresses the need for proper light stimulation, air movement, and development of the immune system.

Pasturing turkeys also exposes the flock to additional challenges, such as:

- Lack of climate control
- Potential soil-borne diseases
- Increased predation threat

It requires more land on which to keep the turkeys and more management of the birds themselves (in terms of moving or catching the turkeys).

— the Standard bred or heritage turkeys. Again, using our car analogy, these turkeys are more like a 4×4 truck. On an open, paved road the truck cannot go as fast as the Corvette, but it can go up and down muddy hills and drive faster than a Corvette over fields. So long as there are few stresses and disease challenges, the industrial turkeys will do fine in this system (like a Corvette on a flat field). But as stresses stack up, the heritage birds, with their superior immune systems, will continue to thrive.

Fixed Buildings with Outdoor Access

Combining fixed buildings with outdoor access may sometimes be necessary to fit your given circumstances. In such a situation, it is best to keep the bedding in the building clean and dry.

TURKEY MANAGEMENT: SANITATION

To avoid soil-borne diseases the turkey grower should provide:

- Land not used by poultry for 2 years or more.
- Land not contaminated by drainage water or poultry manure.
- Land as far from previous range as possible.
- Land well drained and free from swampy areas.
- A range with clover, alfalfa, or other good pasture; or confinement facilities.
- Litter changed often enough to keep the floor clean, dry, and warm.
- Wire platforms (such as sun porches) and floors thoroughly cleaned once a month, or more often if flies are numerous.
- Clean feed and water.
- All feeders and waterers constructed so as to exclude droppings.
- All feed and water pans on wire platforms or on wire floors if practical.
- A dry area around water pans at all times.
- Outdoor hoppers moved to a clean spot on range each week.
- Starting poults on dry mash to simplify sanitation (as opposed to wet mash).

You can manage outdoor access in one of several ways and still have great success.

You can treat the outdoor access as a composting area, applying yard waste, leaves, straw, or some other carbon-rich material. This will create an environment friendly to organisms that will break down waste products and compete with any disease-causing agents.

You can segment the outdoor access into different paddocks, so that each may be given a period of rest and even be replanted to promote plant diversity.

Lastly, you can build sun porches. These are platforms enclosed by wire, with a wire floor, that prevent the turkeys from coming into contact with the soil (and with their droppings). Turkey droppings will accumulate

CLEAN MANAGEMENT

- Old and young stock should be kept separated at all times, and turkeys should be raised away from all other poultry.
- Avoid crowding, chilling, and overheating.
- Prevent cross-drainage and transfer of contamination by visitors, animals, vehicles, and so on.
- Encourage early roosting.
- Allow ¾ to 1 square foot (0.07–0.09 sq m) of space in the brooder house per poult.
- More than 150 poults in one brooder unduly increases the disease hazard.
- Provide range, wire platforms, or small clean yards in front of the brooder house.
- Ample shade on range is essential, avoiding dense shade.

DISINFECTANT

Fortunately for the turkey raiser, lye, which is readily available and inexpensive, has been found quite effective against parasite eggs and disease germs.

Research has shown that a mix of 1 ounce (29.5 ml) to 1 gallon (3.8 L) of water gives good results.

and must be removed. The wind will help manage ammonia levels, but will not prevent exposure of ammonia to the turkeys. The combination of fixed housing and outdoor access can be appropriate to either industrial or heritage turkeys.

The Challenges of Overstocking

Pastured production should not be confused with keeping turkeys crowded into a small pen with mud-soaked ground long bare of living plants. Such an area imitates the industrial model of production, where large numbers of animals are concentrated in a small space, constantly exposed to their own excrement, but to that we have added wet bedding and conditions ripe to grow disease-causing agents. Is it any wonder that turkeys exposed to this muddy, septic material are prone to health issues and disease? If keeping turkeys under such nasty conditions is your idea of pasturing, let me suggest you do what industry has done — put a roof over it!

Overstocking causes health challenges in livestock and poultry largely due to exposure to excrement, damage to lungs, and conditions that favor disease and disease-causing agents. When you understand that a body assimilates nutrients and expels excess and toxins, it is clear how being

Turkeys will be happiest when they have access to pasture and are not overcrowded. They will move about, run, jump, play, and strut, expressing their joy in life.

situated in close proximity to large concentrations of toxins can cause disease. A friend who is a Civil War buff once told me that a significant percentage of Civil War soldiers died of dysentery/cholera due to latrine location relative to the fresh water supply. So, only as far back as 1863, little consideration was being given to the basic biological fact that what goes in and what comes out should not be mingled.

During overstocking conditions, bacteria in excrement becomes airborne, clogging nasal passageways and eventually reducing lung functionality. Ammonia from fresh manure evaporates and mixes with the air the turkeys breathe, damaging lung tissue. Under wet conditions, bacteria and septic material from excrement become encrusted on the feet, feathers, and beaks of the birds, and become a challenge for the digestive tract and a stress factor for the immune system.

Effects of Inhaling Ammonia

To understand the severity of overstocking challenges, let's look at the effects of ammonia breathed by confined turkeys.

At 10 parts per million, ammonia causes damage to the turkeys' air sacs after several weeks of exposure.

At 25 parts per million, damage to the air sacs and lungs will occur within 48 hours of exposure.

At 50 parts per million, the human nose is finally able to detect the presence of ammonia.

After 1 week of exposure at 50 parts per million, damage to both air sacs and lungs will be significant.

At 100 parts per million, some mortality will begin to occur in the flock of turkeys.

Overstocking is hard on the land, too. The deposition of nutrients can be very beneficial to soil fertility at low to moderate levels, but when an excessive amount of any given nutrient is deposited on land, it will act as a poison instead of a benefit. This is not so hard to understand. Just as we all need water, so are proper levels of nutrients good for the land. If the water is increased, it can be dangerous.

Learning from Nature

All agricultural uses of turkeys involve stocking the birds at a much denser rate than Nature had ever intended and then using management to stave off the ill effects of those rates. (This is essentially true for all livestock species.) Dense stocking rates tend to use up natural food resources and cause an overdeposition of excrement.

The only examples Nature gives us of anything similar are movements of large herds of animals and flocks of birds, such as bison crossing the Great Plains of the American West or the annual migration of Canadian geese. In the natural dense stocking rate, the number of animals per acre is conditioned by the element of time — the animals are in the location heavily for only a short period of time and then move on to a fresh area. This is very important, and it modifies the effects of dense stocking rates.

Grazing and Trampling

One of the great benefits of pasturing turkeys is that we are developing a symbiotic relationship between the land and the animals. The pasture benefits the turkeys with fresh, clean air; green foods; insects as a protein source; exercise; clean surroundings; and sunlight. The turkeys benefit the pasture by spreading around their nutrient-rich manure, controlling insect populations, and grazing and trampling plants.

Grazing and trampling help shape an ecosystem. Trampling is important to plant diversity. Animals tend to graze the plants that taste best to them. Given unlimited exposure to a pasture, the animals would graze their favorite plants to death — leaving only those they do not like. If animals are on a pasture densely, for a short period of time, then they consume what they like and trample what they do not.

The end effect is all plants are given an equal stress and then an equal opportunity to regrow. With 45 to 60 days of rest, all plants in that pasture are able to benefit from fresh manure and rest. Over years this allows all plants in that field to mature — developing larger root systems, which hold more moisture and harbor more microorganisms, which in turn better break down manure that helps feed the plants.

A truly sustainable farm integrates all elements to the benefit of the whole. Pasture becomes pen and food source. Manure becomes fertilizer. Exercise becomes trampling which prevents ungrazed plants from taking over the pasture. A sustainable farm, regardless of size of its acreage, is a small ecosystem. (If you wish to learn more about how grazing and trampling benefit a pasture, check resources for Holistic Management International.)

Rotational Grazing

Dr. W. A. Billings was perhaps the first to advocate rotational grazing of turkeys, almost 100 years ago. He noticed that young turkeys, about two to three months old, would closely eat weeds and grass if pastured in large numbers for the area fenced. By moving the fence as soon as the pasture was beaten down, the diets of the young poults were augmented

BEAK TIPPING

During the early 1900s it was common for large turkey producers to practice beak tipping to help prevent feather pecking, eye pecking, and cannibalism. By cutting off the first quarter to third of the top mandible of the beak, producers solved a problem resulting from poor management — that of overcrowding.

Beak tipping can be done with a sharp knife, clippers, or pruning shears.

such that they ate less purchased feed and grew to finish at a slightly better weight. He also advocated leaving that pasture empty for a year.

We now know it is best to wait 3 years before rotating turkeys back to the same pasture, but only 45 to 60 days before rotating in other livestock. Manure from turkeys deposited on pasture will immediately benefit the soil microbes and plants. But only half the organic matter breaks down and is available the first year; the rest is available in the next season. Also, most disease organisms cannot live long in the soil. Those that do will often break down after a couple of years of exposure to the seasons and the elements.

The Minnesota Plan

The basic plan of moving turkeys weekly to fresh pasture is not new; in fact, it's relatively old. Once called the "Minnesota plan," the premise is to control grazing and keep the turkeys on pasture and moving — limiting exposure to their own manure while maximizing the benefits of fresh pasture. This system is good for the pasture and for the turkeys. It requires a good bit of land, so the turkey enterprise must only be a part of the whole farming scheme. The self-applied turkey droppings will fertilize the soil so that subsequent pasture or crops grow well.

Variations on the Minnesota plan have been tried — including leaving the turkeys on one field for 2 to 6 weeks at a time. The soil type, drainage, and the size of your farm will dictate the exact frequency of movement. But suffice it to say that frequent moves spread the manure evenly across the pastures, keep the turkeys grazing clean new pasture, and give the pasture plants time to regrow. The best soil condition is that of well-drained, even sandy, soils. However, turkeys will do well on all but soils that have standing water or which are swampy; such is simply, in combination with the turkey droppings, a breeding ground for pathogens (and parasites). One of the big advantages of rotating turkeys on pasture is sanitation — be sure to keep this in mind when appraising potential pasture.

Fencing Basics

When designing buildings and fencing to protect and contain turkeys, if we satisfy their needs and account for their natural tendencies we will have great success. For instance, young turkey poults are shy and need protection from the elements while they grow and develop their immune systems. Adolescent turkeys are curious, lightweight, and have a developing three-dimensional view of the world (think fly upward). Adult turkey hens need seclusion for nesting, range to explore, and a place to roost. All need fencing.

The best fencing height for adult turkeys is 6 feet, but we must be careful not to have rails along the top, as this will encourage the turkeys to fly up and over. Four-foot-high fencing can also work. Light fencing can be placed at the top of gates to discourage landing on these and going up and over the gate. As a rule, turkeys can fly as high as 9 feet to reach the top of a building. Clipping the feathers of one wing of each of the adult hens, but not the toms, will prevent them from leaving the enclosure and yet still be able to fly onto the roost at night.

As they pass the 14- to 16-week age mark, and from there on, the female turkeys will be the biggest culprits in flying over fences. Males, especially as they age, will tend to stay near the female turkeys. For these two reasons it is recommended to keep the wing feathers clipped on the female turkeys. Another aspect of turkey nature is that they like to take advantage of high ground and will tend to launch themselves downhill while attempting to fly. In enclosures on hilly ground the result is that 9-foot fences at the lower ends of pens may not be high enough to keep in female turkeys, if they have not had their wing feathers clipped.

Besides "wing clipping," antifly devices were used historically to prevent turkeys from flying up onto buildings, and over gates. These simply consisted of 2- to 3-foot vertical poles, usually of wood, with wire strung between to create a barrier. Antifly devices had no top rail, which could have encouraged the turkeys to fly up to roost, and were so designed to give an overall barrier height of no less than 9 feet from ground level.

For those turkey keepers who wish to keep the wing feathers intact on all their turkeys, the higher the fencing the better will be your results. But good management of your turkeys can balance out any shortcomings of your exact facilities. Since young poults molt frequently, and since they are less likely to fly out of an enclosure if they do not realize they can, I would still recommend clipping the feathers of one wing when the birds are first introduced to pasture.

Wing Feather Clipping

Clipping the primary feathers of one wing of turkeys is often done to keep the turkeys contained within their designated yards and off the neighbor's property. Wing clipping unbalances the birds, thus preventing flight. Its drawbacks are that it affects the beauty of the bird, alters the bird's ability to ascend and descend roosts, and handicaps the bird's ability to evade predation and aggression. Wing clipping is comparable to the cutting of hair and causes no pain to the birds. Simply use a sharp pair of pruning clippers or shears to cut the primary feathers of one wing about 1 to 2 inches from the wing itself. Some lightweight individuals may still escape the yard; in such cases the secondary feathers may also be clipped.

Often, it is advisable to clip the wings of young poults entering pasture to prevent them from leaving the safety of their fenced perimeter. Since poults will go through several sets of feathers as they mature,

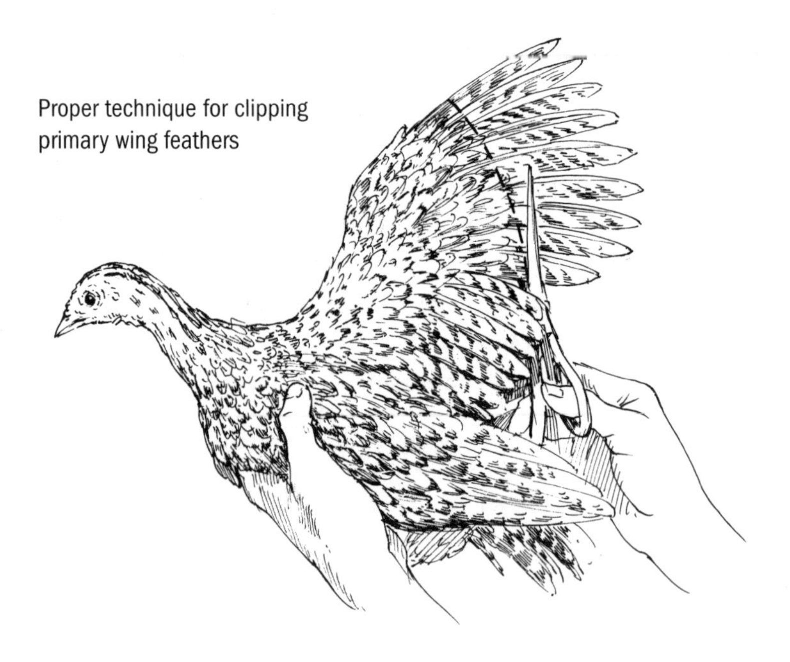

Proper technique for clipping primary wing feathers

clipping as they enter pasture not only helps keep these young birds safe but also will help accustom them to staying within the fenced perimeter. For Standard turkey varieties, clipping should be done at 10 to 12 weeks of age and again at 16 to 20 weeks of age. The first clipping should include all individuals, but the second clipping need only focus on females and light-weight males. With small varieties and wild turkeys, clipping should be done at 7 to 10 weeks of age, at 13 to 16 weeks of age, and again at 19 to 22 weeks of age. Wing clipping times well with movement to a new pasture — catch the turkey in the old pen, clip the wing feathers, and then place it in the new pen, one-by-one until they are all done.

There are instances when wing clipping should be avoided. Turkeys intended to be shown at fairs or poultry shows should not have their wings clipped — though clipping when they are 10 to 12 weeks old will have no negative effect by the time they have matured. Mature, breeding toms should never have their wings clipped, as doing so unbalances the tom and can impact fertility rates. Breeding hens may have their wings clipped, but doing so is undesirable when the hens are intended to be allowed to naturally incubate and brood their own offspring.

Electric Net Fencing

Frequent movement of turkeys requires some form of fencing. Electric net fencing is ideal for use with the Minnesota plan/rotational grazing. It allows for relatively low infrastructure cost — the fencing moves with the turkeys, thus no need to fence the entire property — and weekly moves are easily accomplished due to its portability. This form of fencing, if used correctly, will deter most predators and will safely contain the growing turkeys. This type of fencing is being successfully used in turkey production, and it is well worth your consideration.

When setting up electric net fencing, the best design is to create a rather large square, or rectangular, area with roosts close to the center of the pen. Feeders and waterers should be spread out around the pen, so that no bossy turkey can prevent the others from eating or drinking. Plan to move the netting about once a week — with the idea of keeping the turkeys on clean pasture, preventing the turkeys from overtrampling the pasture, and preventing formation of bare, dirty, or muddy areas.

Rotate more often if the effects of the turkeys are causing bare sections to appear.

You should have extra electric net fencing so that you may set up the fence for the new pen, adjoining the current pen, and simply open one side — in this way the turkeys are always safely contained. Ideally, the feeders should be placed closer to the side to which turkeys will rotate; it will be easier to move the turkeys by opening the fence and moving the feeders — the turkeys will follow.

Electrifying

Before setting up a new fence, it is a good idea to mow a 4- to 6-foot-wide strip where you will stake the fencing into the ground. This will prevent pasture plants from grounding out the electric of the fence — taking away the shock that discourages turkeys, as well as predators, from challenging the fence. You want the fence tight and not sagging. A good idea is to place a metal T-post outside each corner, and use baling twine, or other nonconductive string, to draw the fence tight. Be sure the fence does not touch the metal post, or it will ground out and will not provide any shock.

Chargers can be solar or plug-in — either will work well. The real key to having a hot fence is to provide a good ground. If you are using a plug-in fence charger, and running a line out to the fence, use two to four 8-foot-tall grounding rods and hammer them into the ground, leaving only 6 or so inches above ground. Connect each of these with wire or grounding cable, and the charger will have a strong shock. When using a portable solar charger, you need a good ground as well. I like to use a solar charger that mounts on a metal T-post. I like to use the T-post as the ground, connecting the ground wire to the post. And during dry parts of the year, I will connect it with grounding cable to one or more of the other metal T-posts to increase its grounding ability. When you come to do your daily chores, you should empty the waterers on the ground near the grounding post. By making this soil damp, the grounding ability becomes more effective. The charger's solar panel should face south and be in full sun; if possible, it should be located conveniently for turning on and off while visiting the turkeys.

Movable Roosts and Feeders

The roost is placed in the center of the enclosed pasture to prevent clever predators, such as foxes and coyotes, from scaring the turkeys off their roosts at night. In this way if some of the turkeys fly off their roosts in the dark, they are likely to land inside of the enclosure and not outside of it. You see, foxes and coyotes are smart enough to work in groups to scare the turkeys and then trap the ones that make it over the fence. These predators seem to understand that the loose turkeys will try to force their way back into the pen, but will be trapped against the fence instead. One farm I know of lost 26 turkeys in a single night — the foxes having worked together on opposite ends of the fenced pen, one side barking and the other quietly waiting for the scared turkeys to land outside the pen where they became easy prey as they turned back to try to reenter their pen. Remember, turkeys are designed in shape with a small head on a relatively long neck, and then a cylindrical body. They use their heads to find openings in brush and then force their cylindrical

Set up the roost in the center of the pasture to make it harder for predators to sneak up on the birds.

bodies through with their powerful legs. Fencing creates a barrier that allows their heads through, but not their bodies.

Feeder and waterer design and placement are of high importance. Feeders and waterers should be spread around to reduce competition. But they should also be seen as a tool to help control how well the poultry will use their little bit of pasture. If poultry will be left in the same location for 2 or more days, then each day you should move the feeders and waterers a little. This will spread around any spilled grains, keep the birds moving, and help to teach the birds to forage larger areas. Waterers should be so placed as to prevent creating muddy areas around them and should be designed so that the turkeys cannot walk in them or foul them.

Feeders should be designed to protect the feed from rain, preventing the accumulation of moldy feed. They should have watertight lids or roofs that should be designed such that driving rain from a windy storm does not dampen the feed. Feeders may also attract wildlife, such as rodents and birds. Some feeder designs allow them to be closed at night to prevent becoming a source of food for rodents. Moving the feeders daily will accomplish much the same effect.

Permanent Fencing

In some situations, such as with multispecies grazing or multispecies pasture rotation, permanent fencing will make the best sense. In such situations woven wire fencing of the type for horses, with openings 2 inches wide by 4 inches high (5 × 10 cm); turkey wire, a welded wire also with openings 2 inches wide by 4 inches high; or even American woven wire, with very close horizontal weaves for the first 2 feet (60 cm) and wider openings near the top, will all produce satisfactory results (the first two types being preferred). It would still be an excellent caution to run an electrified wire along the outside, or along the outside and inside of the fence. The electric wire should be attached 6 inches (15 cm) from the ground and will greatly deter predators. A second strand run at 12 inches (30 cm) above the ground will greatly increase the level of protection and is worth the extra cost.

Modern metal fence posts with insulators will work as well as pressure-treated posts. Brace corners properly and stretch the fence to prevent sagging. Low spots are of concern and should be addressed while the fence is being run. Some poultry keepers, with smaller enclosures, will bury wire to prevent predators from digging under fencing. For larger areas this may not be practical. The use of livestock guardian dogs outside the pen, or inside when properly conditioned to live peacefully with the turkeys, as well as the electric line at 6 inches should deter predation.

Permanent fencing may be used in a variety of ways. It can be used in conjunction with portable electric net fencing, providing one, two, or three sides of your fence. It can also be the sole fence used to contain your turkeys. When immobile, permanent fencing is used to contain your turkey flock, it becomes very important to practice good sanitation and pasture management. Heavily soiled land should be limed annually —with the turkeys removed for 4 to 6 weeks. Pasture should be reseeded annually to help cover bare areas. When outdoor feeders and waterers are being used, they should be moved frequently and should be set up on wire framed boards or slats to prevent the turkeys coming into contact with muddy areas or moldy feed. See chapter 8 for more on fencing.

Guardian Dogs

Livestock guardian dogs are a great asset. There are many breeds to choose from, and one may find a mixed breed or dog of another breed that will work too. Livestock guardian dogs are effective because of their presence. Predators know that if they get hurt fighting something, they may not be able to hunt anymore; so most will avoid a confrontation if possible. Livestock guardian dogs will chase and drive some predators away, but their main function is to patrol and mark their territory, and to bark — especially at night. A lazy looking, big, white dog that sleeps most of the day may be the best protector your poultry has ever had. Such a dog is usually up all night, patrols its territory, barks at anything that moves, even fierce creatures like deer and bunnies, and marks the territory by pooping and peeing along fence lines.

Guardian dogs need little to no training. If your farm has frequent visitors, then it is well to introduce the dog to many people starting from the time it is a puppy (socialize). If you have few visitors and a concern that a two-footed predator may sneak in to steal turkeys, then do not socialize the dog—but be warned that if the dog is not socialized, the chances of it biting someone do increase. I like to train my livestock guardian dogs to sit, to "go to the barn," and to "drop" anything in their mouth. These dogs are by nature very independent, so when I say "sit" the dog will choose its own spot to obey. They consider me alpha dog, but that doesn't mean they think I rule them.

Never beat such a dog when training or correcting. Instead, first try a sharp "no." In most cases your dog will look sad, as if you just smacked him. If you must correct him more strongly, then grab the scruff of his neck with one hand, hold for a second, and then release. That is a pretty severe correction in his mind. The strongest correction is when you grab him by the throat and make him lie on his back. Loosen your grip a little and he will part his legs and tuck his tail—that is the most sincere "I am sorry" you can get from the dog. Let go and pretend nothing ever happened, as the dog is sorry and has no way to express this any more strongly.

The best livestock guardian dogs are those that can focus on their jobs. For this reason, it is best to have spayed or neutered dogs. It is also a good idea to never take such a dog for a walk off property. Why? Doing so is teaching the dog that this area is part of her property too. She will attempt to patrol this additional, expanded section of her property, even trying to drive away animals that she deems do not belong, such as the neighbors' visitor's poodle. . . .

Guardian dogs come with a price — not only monetary but also the time and effort they require when they are young. For the first year or two, depending on the dog, playful puppy behavior may be a problem; after all, we don't want the dog chasing, grabbing, or killing our turkeys. The best way to train the dog to see the turkeys as his charges is to start by keeping him outside of the turkey pen. Let the dog come in with you as you work, but never leave him unsupervised. If he tries to play, remove the dog and try again the next day. You may need to correct the dog a few times. My current dog is a natural with poultry. From a puppy, she never tried to play with or harm a chicken or turkey. The one before, who was a great protector, could not be trusted alone with poultry until she was two years old. You have to work with them and observe their behavior to know when they are ready.

Lastly, remember that livestock guardian dogs work cheap, but they do so because they see their charges as defenseless members of the pack. So if you plan to harvest all your turkeys, consider that the dogs need one or two to stay with all winter long or they will feel lost. See chapter 8 for more on guardian dogs.

Roosts

Turkeys like to roost. They prefer to do this in areas with good ventilation and light, and so you may experience difficulty in getting turkeys to roost if indoor roosts are placed in a dark, tight corner. When planning roosts, allow a minimum of one linear foot of roost space for each turkey. Stepladder-type roosts work well, with each succeeding roost a little higher than the first. The lowest roost should be about 20 inches (0.5 m) off the floor and the roosts should be 2 feet (0.6 m) apart, on center, with the top one at least 18 inches (45 cm) from a wall (when used in or against a building).

When roosts are the primary method of preventing nighttime predation, position them approximately 7 feet (2.5 m) off the ground. To make it easier for the turkeys to access such high roosts, you can set up

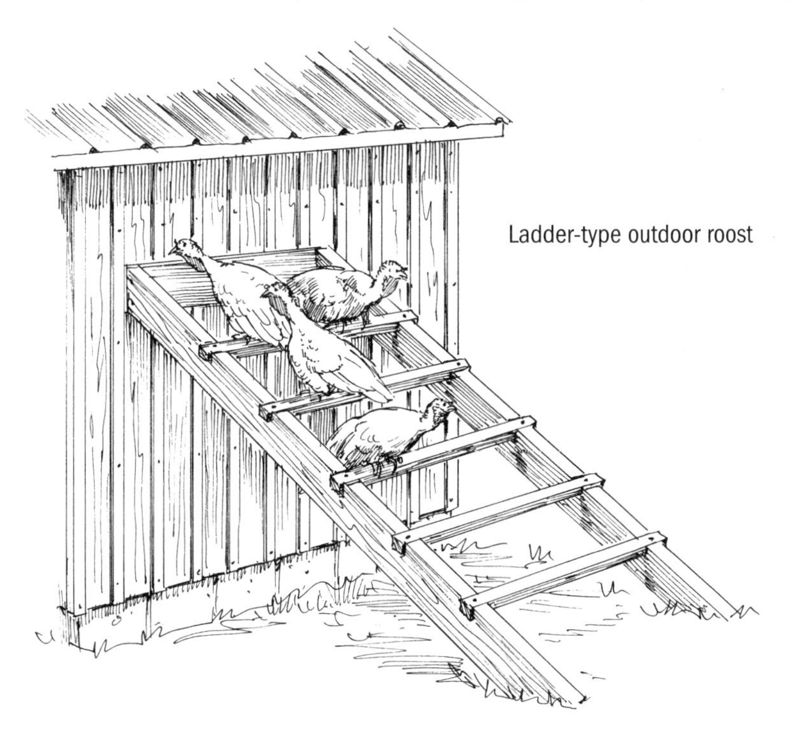

Ladder-type outdoor roost

Freestanding roost

gangplanks. Industrial, broad-breasted turkeys tend to roost on or near the ground — when planning evening quarters you should keep this in mind to prevent losses from predation. (See chapter 3 for more on range shelters, and chapter 8 for more on protecting against predators.)

You can build a roost from 2×4 boards (above), positioning them with the 4-inch (10 cm) surface horizontal. This gives a nice flat surface that will not damage the soft breastbones of the growing turkcys. Another advantage of turning the boards this way is that as the turkeys roost, their breasts will cover their feet — keeping toes from suffering frostbite during winter.

Elevated antipredator roost

Shade

One of the biggest challenges with pasture grow-out of turkeys, shade is too often overlooked during the planning phase of the enterprise. Turkeys of all ages will enjoy strolling out in the sunshine, but intense summer sun combined with high temperatures can lead to heatstroke and catastrophic losses. As you plan for a solution to providing shade, it is well to think of some human equivalents.

How long ago was it that every home had a front porch and shade trees? Today we "plant" houses in fields without trees and attach central air conditioning units. If you have ever sat on a rocking chair or swung on a front porch, or sat in a lawn chair under the family's favorite shade tree, you will remember that this area offered not only protection from the sun but always had a nice breeze blowing. You are your own best guinea pig; spend some time during a hot summer's midday under your favorite shade tree, and learn the feel of both the lower temperature and of the slight breeze. Having done so, you will be better able to appraise the merit of any shade structure or source you intend to provide for your turkeys.

When providing shade for turkeys we need to think of five main points: effective cover from the sun; air movement and quality; space provided relative to the size of the flock; impact to the soil; and disease control. Obviously we want to block or reduce the amount of direct sunlight, but a shade structure with poor air movement may still overheat the birds by failing to whisk away their body heat and the moisture they expel while exhaling. Likewise, a structure that is not large enough for the entire turkey herd will cause crowding. When we leave the structure in place for long periods of time, or fail to rotate the turkey herd across the pasture, the constant trampling by the turkeys will kill the pasture plants and create bare areas with a hardpan of soil.

Bare areas frequented by the turkeys can become heavily manured and pose some disease potential. All living beings develop disease when constantly exposed to their own manure; again, think of cholera and Civil War armies as an example. One of the great advantages of pasture is that the animals are removed from concentrations of manure.

A tarp turns an outdoor roost into a shady refuge.

Trees and bushes may provide some shade, but to utilize prime pasture, you must provide structures for shade. Consider the following:

- Roofed turkey roosts can be a portable shade source.
- Portable shade can be created using lightweight, framed structures with tarps, tin, shade cloth, or even burlap roofs.
- You can grow corn or sunflower strips or patches.
- Fence rows, when overgrown, provide excellent shade for turkeys.

If you are using portable roosts with or without a roof, there is a simple way to add shaded areas to your pasture — attach a tarp or shade cloth to the framework of the roost and use metal fence posts to secure the other two corners. You can attach the tarp to the left, right, or rear of the roost framework. The shaded area should keep the turkeys a safe distance from the fence enclosing them, and yet provide the best protection from the sun. This will give you a nice covered area that is easy to move. As you move the frame across your pasture, simply pull up the two posts, drape the tarp over the framework, and pull the roost to its new location.

Pasture

A variety of pasture plants work well for turkeys. Alfalfa is an excellent forage for turkeys. Other good crops include grasses, like orchard grass, vetch, and various clovers. A mixed pasture is to be preferred and overseeding it with some mustard greens will help reduce or prevent parasites. (Mustard greens are high in the mineral sulfur and thus discourage many parasites—both internal and external.) Chickweed should be encouraged, as well as dandelions, and both are excellent greens to feed to poults while they are still in the brooder.

Pasture amendments should be applied no sooner than 4 to 6 weeks before turkeys are introduced. It is an excellent idea to lime the fields as turkeys exit, to help balance the acid level of their manure. Seeding should happen long in advance of the turkeys, or seedlings will have little chance to mature. Seeding can follow the turkeys' exit. Interestingly, Landino clover seeds in the ground will tend to germinate in areas left bare by the turkeys.

It is generally advised to stock no more than 100 turkeys per acre. Exact stocking rate must consider the amount of time the turkeys will spend on any given section of pasture, as well as the quality of the pasture and the amount of plant growth it supports. Obviously, a pasture in Colorado may not support as many turkeys as one in Illinois. This means that 1,000 turkeys will require about 10 acres of land. So be sure to consider the amount of land you have available when deciding how many turkeys to raise. If you wish to grow your own turkey feed as well, it will take about 17 acres to grow the feed to grow and finish that same 1,000 turkeys.

Sandy soils can stock up to 1,000 turkeys per acre and clay soils up to 300. Stocking at such a high rate will be hard on pasture and an overly high level of nutrients will be deposited onto the pasture. So if you need to stock at such a high rate, plan to use a different location for the next few years.

HARDEN OFF

by Frank R. Reese, Jr.

The growers used to always say "harden off" your poults before putting them on range or before turning all the hoods or heaters off. "Harden off" meant to make sure the heads of the birds were fully mature (loss of all baby fluff), birds were in their second feather growth, and they had been exposed to cold and heat outside. This meant the birds were better equipped to handle weather changes.

Manure

Turkey manure is an excellent source of fertilizer. When you are raising the turkeys on pasture, they will self-apply their manure, spreading it around the field. Since they deposit about one-third of their manure while on the roost, moving the roosts a little each day will help prevent an overly heavy concentration in one area of the field. Fresh turkey manure is organically rich and only 50 percent of the nutrients are released the first season — so the benefit of this natural application can be seen for two or more seasons. Interestingly, turkey manure will help naturally break down the limestone of the soil, as rain on turkey manure will form carbonic acid.

In their book, *Day Range Poultry*, Andy Lee and Patricia Foreman give some good information on turkey manure. They say turkey manure is high in nitrogen (N), phosphorus (P), potassium (K), and trace elements. They give turkey manure a value of roughly 10–10–10; and say that in 8 to 12 weeks, 100 turkeys will deposit about 240 pounds of nitrogen per acre.

The American Society of Agricultural Engineers offers some interesting information on turkey manure nutrient levels, as measured from confined turkeys, in their 2005 publication, *ASAE Standards, Manure Production Characteristics*. They assume a start-to-finish time for male turkeys of 133 days and female turkeys of 105 days (values that indicate industrial turkey stock). Nutrients are measured as total volume produced, and this value is given for males and females separately. This is an interesting angle on measuring nutrients, valuable to help explain the value of the manure as found in litter. If we assume a flock of 50 percent male and 50 percent female turkeys, we can expect them to produce, on average per bird: 58 pounds of manure consisting of 0.89 pounds of N, 0.26 pounds of P, 0.41 pounds of K, and 17 pounds of calcium (Ca) with a total moisture content of 74 percent.

Alex and Betsy Hitt are farmers I had the privilege to meet and interact with when I lived in North Carolina and worked for the American Livestock Breeds Conservancy (ALBC). The Hitts were produce farmers and they worked with ALBC as we strove to bring heritage turkeys back to the marketplace. Though produce farmers, the Hitts quickly discerned the most important aspects of turkey husbandry when they integrated turkeys as a part of their farm and their offerings for their consumers. They valued the turkeys for two reasons — the self-applied manure and the bug-eating ability of turkeys. They figured the manure deposition at a rate of about 2 tons per acre per year, which works out to a flock of approximately 900 heritage turkeys or 500 industrial turkeys. The amount of manure is roughly equal to the amount of feed.

As you can see, there are a number of ways to compute the value of turkey manure. But it should also be clear that turkey manure offers some great values. If the turkeys are processed and the enterprise only breaks even, then the value of their manure, self-applied, clearly offers a profitable benefit.

Putting Poults on Pasture

In order to ensure our turkeys grow to their full potential, we need to do everything we can to safeguard their health while they are very young. Young turkeys are slow to develop their immune systems. They

TURKEY MANURE NUTRIENT LEVELS

(All values per finished animal)

Nutrient	Values in lbs Turkey (male/female)	Values in kg Turkey (male/female)
Total Solids	20/9.8	9.2/4.4
Volatile Solids	16/7.8	7.4/3.5
Chemical Oxygen Demand	19/8.8	8.5/4.0
Biochemical Oxygen Demand	5.2/2.4	2.4/1.1
Nitrogen	1.2/0.57	0.55/0.26
Phosphorus	0.36/0.16	0.16/0.074
Potassium	0.57/0.25	0.26/0.11
Calcium	17/17	7.7/7.7
Total Manure	78/38	36/17
Litter	–	36/17
Moisture	74%	74%

Volatile Solids = Total organic matter per animal finished

Source: *ASAE Standards, Manure Production Characteristics*, 2005 publication, American Society of Agricultural Engineers

MOVING POULTS TO RANGE

Moving poults from the brooder house to the range can be quite a shock for them. To lessen this, open the doors and provide a range shelter at eight to ten weeks of age. The turkeys will naturally move to the roost of their own accord.

are vulnerable to disease, and multiple stress factors can be very hard on them. (See Stress Factors on page 99.)

All life forms are equipped with immune systems to fend off various pathogens. When a body is stressed, part of its defenses are occupied and so a second stress factor has more likelihood of impacting it. Adult turkeys are incredibly hardy, but for the first few weeks of a poult's life we must guard its health by limiting exposure to pathogens and stress factors.

It is considered good management to keep young turkeys off the ground until they are six to ten weeks old. One method that was once very popular is to build sun porches — wire platforms, raised off the ground, with wire floor, walls, and ceiling — and to use these to give the young birds outdoor access without exposure to pathogens possibly found in the soil. Sun porches are predator proof and greatly expand the amount of space per poult, preventing feelings of being overcrowded. Poults should have their first full set of feathers before being let out onto the sun porch — they should be roughly four weeks old or older. Turkey poults raised on sun porches are usually moved to pasture around 10 to 12 weeks of age. (See page 106 for more on sun porches.)

We could simply keep our turkeys in buildings until they have "hardened off." If you are using a building, and if it has plenty of room, it is not a bad idea to keep the poults inside until they are at least six to seven weeks old. This gives the poults a chance to gain a little size, which gives them more mass — thus less likelihood of chilling — and a chance to develop more fully. At around six weeks of age they reach a stage where the down has come off their heads and their second, stronger set of feathers is in to protect them from the elements. Poults should have molted into their second set of feathers before they are exposed to rain or snow.

NATURE AND BOREDOM

Whether using a sun porch or letting the turkey poults outside once they have hardened off, we must consider how the poults will interact and how nature will cause them to react to the "environment" we have created. In nature; a typical brood size is up to 16 turkey poults. Modern methods bring together 150 or so turkey poults. Turkeys have an inquisitive nature; having a large group in a small space is a surefire recipe to cause feather pecking, cannibalism, and pileups.

Turkeys like to gather in a group. When two or more turkey groups are near each other, they will try to join each other.

Buildings should have multiple doorways and these should be large enough for several turkeys to enter or exit at the same time. Do not put poults in a position in which one bossy poult can block a door and keep many others from eating, drinking, or returning to the brooder.

Turkeys like to roost. It is embedded in their nature. Roosts can reduce the feeling of being crowded by allowing some birds to be up, off ground level.

Sight barriers also help reduce the feeling of being crowded. Sight barriers are any solid device that blocks the view of one turkey to another. Some feeders and waterers are large enough to act as sight barriers. Roosts and tarps can also be placed so as to block the view of other turkeys. Sight barriers not only reduce the number of turkeys seen at any given moment, they help a turkey get away from the attention of another that is attempting to bully it — thus sight barriers can reduce stress.

Remember, frightened turkeys will pile up. Pileups will result in the death of some turkeys and damage to others. We can largely prevent pileups by moving turkeys slowly and by handling them in a gentle fashion. Clint Grimes once showed me that it is easier to catch a turkey by walking than by running — they let you get closer. Turkeys will run if you run.

Always keep the nature of a flock of turkeys in mind as you move and work with your turkey flock.

Introducing Poults to Pasture

The best way to introduce the poults to pasture is to fence an area adjoining the building the poults are in, and open a door to give them access. In this way the poults still have the security of their brooder house and the warmth of the brooder available as needed. You may be surprised how reluctant to venture outside the poults will be for the first day or two. On day 2, place half of their waterers and feeders outside to help encourage them to explore. On day 3 or 4, place a movable outdoor roost into this fenced area.

By day 5, most of the poults should be spending considerable time outside; move the remaining feeders and waterers outside and from this point on, move some of them a little farther out each day. By day 5 you will begin to see poults roosting at night outside — this is perfectly alright, unless unusually cold and wet weather arrives. If the roost has a roof, and the poults have hardened off, then it is ideal to let them decide when to live outdoors instead of in. Usually within a week or so, all the poults will be sleeping outside — at this point close the door and begin your pasture rotation.

Move the turkeys to new pasture frequently. Such a move is better for the land and for the turkeys. It supplies fresh pasture and insects for the turkeys to use to supplement their diet. It gives them stimulation to satisfy the curiosity of their nature. Plan moves to happen every 4 to 7 days depending on the age of the turkeys, the size of the flock, the size of the area fenced, and how well the turkeys have utilized the area. Better to have medium-sized pens with frequent moves than large pens with infrequent moves. In this way the turkeys will trample the plants they do now eat — giving similar recovery time to all plants in the pasture and thus improving the quality of the pasture over time.

Moving Turkeys

Make the move in the morning. The roosts and waterers should also be moved then, or a few hours later. Moving early in the day means the turkeys have shade and so they can go to roost as dusk arrives. If you allow your turkeys to empty the feeders the night before the move, then in the morning they will be willing to follow a feed bucket quite some distance.

Turkeys may also be driven a considerable distance to a new pasture. This works best with two people — one leading and carrying a bucket of feed, and the other walking behind with a stick (or a flag) in each hand. Walk slowly and use the sticks as extensions of your arms to keep all the turkeys moving in the same direction (see box below). Any injured turkey should be gently carried to a safe location and given time to heal (see illustration on page 2).

HERDING TURKEYS

Herding turkeys is best done by two people. One walks in front, and the turkeys follow. The second person carries two 3-foot-long (1 m) branches, holding them out to his or her side at about waist height, pointed down so that the tips are about a foot off the ground. This person keeps the birds at the rear moving and together — looking to the turkeys like a giant boss tom with his wings lowered. Both people should move slowly, and the second person can reach out with a stick when necessary to drive a would-be straggler.

Turkeys are subject to injuring each other when frightened; pileups may also occur, resulting in death. This is why it is always important to move slowly and keep the turkeys calm when moving them.

Drovers used to travel around and gather up all unwanted livestock and drive them, on foot, to market. It has been 60 or more years since any drover has come around most parts of the country, but the techniques they used are still valid today. Drovers used one person in front of the livestock and another person or dog at the rear. The type of dog used to drove was not the fast, agile Border Collie type, but a large, slow, pondering dog like the Old English Sheep dog. Likewise, if you use a dog for herding your turkeys, it should be nonaggressive to the birds and work slowly.

PASTURE POINTERS

- Poults should not enter the pasture until six to seven weeks old, after they have been hardened off.
- When using sun porches, introduce poults to pasture at an age of ten to twelve weeks.
- Place some feeders and waterers outside to lure the young poults into exploring pasture.
- Doorways should be plentiful or wide enough to allow several turkeys to enter or exit at a time.
- Place an outdoor roost in the pasture and wait for the young turkeys to decide when they are ready to roost on it all night — this is when they are ready to leave the brooder building behind.
- Move the feeders and waterers a little each day, in the direction that you will move the turkeys. In this way you will train the young turkeys to forage and to move across the pasture.
- Clip the wing feathers of young turkeys (and female turkeys) to prevent them from flying out of the pen. Plan to clip wings at ten to twelve weeks of age, and again at sixteen to twenty weeks of age. With small varieties and wild turkeys, clipping should be done at seven to ten weeks of age, at thirteen to sixteen weeks of age, and again at nineteen to twenty-two weeks of age.
- Protect your turkeys from predators by using electric fencing and guardian dogs or livestock.

Avoiding Truck and Trailer Pileups

If you cannot herd your turkeys to pasture then take extra care while using a truck or trailer to move your young turkeys. During this time they can become frightened and will pile at the head of the trailer, causing many to suffocate. If you must use a trailer, here are some tips:

- Use individual turkey crates to move your birds.
- Take the time to notice the sex of each bird, and place only toms together in some crates and only hens in other crates.

- Keep turkey groups limited to approximately 150 or fewer birds. After they are twelve weeks old, groups may be combined to form larger flocks.
- Use visual barriers to reduce feelings of crowding.
- Turkeys like to flock together — keep groups separated by distance.
- Turkeys that get out of a fenced pen will try to force their way back in, toward their flock.
- Herd, catch, or handle turkeys calmly and slowly. Turkeys hurt each other when they panic.
- Move the turkeys and their pen at least once per week (every 3 to 5 days is preferable). This prevents overgrazing the pasture, spreads the manure more evenly, provides the turkeys with more usable forage, and prevents boredom.
- Provide plenty of shade so that turkeys are not lost to heat stroke. There should be enough shade that there is plenty of room left even when all the turkeys are huddled for shade.
- Shaded areas must also have plenty of air movement to wick away the moisture and body heat expelled by the turkeys.
- Bored turkeys will turn their inquisitive nature upon each other. Provide the turkeys with new stimulation by moving them to new pasture or providing objects to peck at, such as bones on a string.

MANURE POWER

A Minnesota Power plant began using turkey droppings for fuel in 2007. The turkey manure, 700,000 tons of it, produces 55 megawatts of power per year. There are three such plants in England too.

- Large birds will pile on top of young birds, even in a crate. Take notice of size and weight of the turkeys you're moving and divide them by size as well.

Turkeys will not pile on top of each other on a roost. This is how Mother Nature prevents pileups in the wild.

◆ ◆ ◆

Anyone can have great success raising turkeys if they remember the nature of the bird and learn to work with it rather than against it. Use the list on page 138 to be sure to address the most important aspects of a turkey's nature and focus on the most important age for various chores.

8

Protection from Predators

EVEN THOUGH TURKEYS ARE quite large birds, many predators see them as an excellent food source. Turkeys are native to the Americas and have evolved side by side with many natural carnivores. Some of these are night hunters — raccoons, opossums, weasels, mink, and skunks. Others hunt both day and night — foxes, coyotes, wolves, dogs, bears, and cougars. If you are unaware of the types of predators in your area, contact your local Department of Wildlife and Game, Parks Department, or local Extension office to learn which predators you may encounter, and any laws associated with these animals.

Small Predators

Rats, cats, snakes, and weasels are all threats if they have access to the brooder or brooder building. Small predators will prey upon turkey poults, especially when the poults are still in the brooder. Cats, though rarely, will sometimes attack poults that are several weeks old. Weasels can be a threat to turkeys of all ages.

Rats

Protection from predators starts in the brooder. Nothing is more disheartening than to find a brood of poults maimed, dead, or dying.

The single most common predator on the small farm or homestead is the rat. Rats have been known to maim young turkey poults, biting off legs or wings, and to drag the still-alive victims down into their burrows to finish off later.

Habits and clues. Rats operate primarily at night. Since they frequent buildings around the farm, often arriving in the late fall after harvest, a good rat prevention program is a very good idea.

Always clean up spilled feed, keep fresh feed in rat-proof containers, and patch holes with metal sheeting, such as can lids. When using traps and poison, be sure to place these away from the turkeys so that they are not harmed by the traps or killed by consuming the poison. (See more about rat control in chapter 12.)

Cats

Domestic cats — whether barn cat, house cat, or feral stray — will sometimes prey upon young poults. While these feline members of the farm ecosystem aid in helping control mice and even rat populations, their access to turkey brooder buildings is best left to times when these buildings are empty.

Habits and clues. Cats may prey upon poults during day or night, and leave little evidence. It is usually best to keep the cats from having access to poults younger than six weeks of age. Keep cats locked out of brooders and brooder buildings.

Snakes

Snakes pose very little risk to adult turkeys; in fact, adult turkeys have been known to encircle snakes and attack them. Snakes will swallow turkey eggs, however, and even young turkey poults.

Habits and clues. Snakes enter pens through holes and gaps in the wire. They like low objects such as nests, corners, boards, anything that provides some cover. Snakes leave little evidence — eggs or poults simply disappear, one or two per day.

An old method of keeping snakes out of barns, from my friend Byran Childress of Virginia, is to pour copper sulfate all the way around the outside of the building — snakes appear to avoid crossing soil thus treated.

Weasels

Nocturnal predators, weasels and mink (and even ferrets) will attack young broods of turkey poults.

Habits and clues. Members of this family will kill the poult by biting near the base of the skull and then feed on the blood. Weasels and mink tend to pile their victims neatly and leave closely spaced caninc marks on the necks, heads often decapitated. They will eat eggs, too, and will completely crush the shells when doing so.

Prevention: Secure the Coop

Rats, cats, snakes, weasels, and mink can all fit through very small openings or holes. Be sure to inspect brooder coops before a new batch of poults arrives, and cover any holes with metal flashing (or the tops of cans). Wire for brooder pens should be 12-gauge galvanized wire mesh with ½-inch (1.2 cm) or ¼-inch (0.6 cm) openings.

Medium-sized Predators

Some predators can climb and even open simple latches. Skunks and opossums typically prey upon eggs and turkey poults, but raccoons can be a threat to turkeys of all ages. Often these medium-sized mammals represent the largest of threats.

Skunks

Skunks will eat young poults and turkey eggs. If your brooder pen is well secured, however, they will pose little threat.

Habits and clues. When skunks attack poults, they typically eat the heads and entrails. They are nocturnal predators and leave an odor as a telltale sign.

Raccoons

Raccoons are very clever predators. Many poultrymen have said that raccoons should be issued poultry-judging licenses as they always seem to prey upon the best birds. They are capable of opening many types of simple latches, they can unravel standard poultry wire, and they will even reach into the pen to pull parts of their victims out.

Raccoons are one of the hardest predators to thwart. They can climb and dig and can even unravel chicken wire.

Habits and clues. When an old sow raccoon is teaching hunting to her young, they will often kill many birds without eating them. When a raccoon does eat a turkey, it will usually eat the head and the breast. Eggs that have been opened at one end and are empty and clean are one sign of raccoon activity. Raccoons hunt almost exclusively at night and will prey upon eggs, poults, young turkeys, and adult turkeys.

Opossums

Opossums are another nocturnal predator. Young poults and eggs are desirable foods for opossums.

Habits and clues. They typically eat the rear end of the birds, and when eating eggs they crunch the shells up, leaving many little broken bits. Opossums rarely attack adult turkeys — but rarely does not mean never.

Prevention

Skunks, raccoons, and opossums can all be thwarted by solid buildings, electrified wire fencing and netting, and tightly meshed, galvanized wire pens. See Fencing Options on page 151 for some strategies I've seen firsthand in my home state of North Carolina.

Large Predators

Moving the young poults out to pasture only increases the types of predators that may have access to the birds. Foxes, coyotes, wolves, and domestic dogs all pose threats to turkey herds. Aerial predators are also now a threat, as are large feline predators such as bobcats and cougars.

Foxes and Coyotes

Foxes and coyotes are both very clever and will use teamwork in attempts to get a nice turkey dinner. It is not unusual for a single fox or coyote to bark or howl to try and spook turkeys into abandoning their roosts at night. When the roost is near a fence, some turkeys may fly and land outside the pen — only to become dinner for the visitor's friends, who wait silently outside the fence on the opposite side from their barking or howling partner. The frightened turkeys will try to force their way back through the fence, using their natural instinct to charge headfirst through brush; in the dark they make easy victims as the wire allows their heads to pass but not their bodies. Very clever are foxes and coyotes.

Habits and clues. There are a few clues that will help you determine whether you have coyotes or foxes. Coyotes tend to roll on the ground near a kill, and may urinate or defecate. Foxes leave few clues, carrying the carcasses to their den. A fox will eat by biting through the area below the ribs; this is one indicator when such kills are found. Both coyotes and foxes tend to grab their victims by the necks.

Seldom seen during daylight, coyotes often work in teams to frighten turkeys into flying out of the protection of their pens.

Foxes consume a large number of rodents and so can be very beneficial to the farm ecosystem. But don't trust them with your turkeys – maintain good fences to keep foxes out.

Wolves

Wolves will attack domestic turkeys on pasture. They range only in specific parts of the United States, so check with your local game officials to see if they are a threat in your area.

Habits and clues. A wolf will usually grab a turkey by the back. A single wolf will eat only a few birds but can kill as many as 50 to 100 in a single night. It will carry off many more and bury them, often returning within a night or two.

Dogs

Domestic dogs, even the neighbors' cute little "Fifi," can become serious predators to turkeys of all ages. Fifi and her friends are related to wolves and can go primitive when exposed to live turkeys whose owners are not around. Dog attacks are terrible. The dogs are not killing for food, but for the joy of chasing the colorful flashing turkeys.

Habits and clues. When dogs attack, they often kill large numbers and may even leave many wounded birds alive. They will bury or partially bury some birds. Bite marks appear randomly placed; Fifi is just biting whatever part of the bird is closest to her. Dogs will lick off some feathers — little eating usually takes place.

Dog attacks can occur in the day or night. Flashy movements and smells draw their attention. Dogs will usually return to the scene of the crime, hoping for a bit more fun. Strangely, the sweetest little dog sometimes shows its wolflike ancestry when exposed to prey animals. They will seek out weak spots in fencing but seldom think to jump over them. They also show little cleverness in disguising their predatory intentions.

Other Large Predators

Bobcats, cougars, and bears can be problems for your flock as well. All three are nocturnal predators — though bears can also be diurnal.

Bobcats can climb wooden fence posts and cross-bracing; they can even jump over low fences. When bobcats feed on turkeys, they usually take the whole bird. They kill by biting heads and necks and leave clean-cut marks.

Cougars are rare and elusive predators, and they usually stalk their prey. Like bobcats, they can climb and leap, and only high fences are an obstacle to them. They rarely kill more than they can eat, unless a female

LEARNING THE HARD WAY

While some dogs will show no aggression to poultry, you may benefit from my own experience. I once had a dog that had to go on hormone treatments when she aged. The result was that I came home to find 17 birds dead in various ways, some buried. After showing no aggression to poultry for 10 years, my dog even attacked a bird right in front of me, and then acted confused when I corrected her. With the veterinarian's blessing, we cut her dosage in half and she went back to being a reliable farm and flock protector.

WHEN DOGS ARE BAD

When you find that a dog has attacked your flock, please keep in mind that under better conditions this may be someone's loving pet. Identify the dog — take a picture. Contain the dog if possible. Next, call your local Animal Control or police and report the incident. In most states the owner of the dog is held responsible. While a problem animal may be seized or put down, often a fair resolution can be found.

is teaching her offspring. Cougars kill by pouncing from behind and usually bite near the base of the neck.

Bears are omnivores and can use their strength to break through most barriers. They are drawn to food sources by their sense of smell. Bears are as apt to play with their food as to consume it; they can damage and kill turkeys with a variety of wounds. Note that the feed is as large an attraction for bears as are the turkeys themselves.

Flying Predators

Electrified netting or fencing can keep out nearly all walking predators. But there are other types of predators that should concern us. How about flying predators?

Hawks, Eagles, and Crows

Hawks, eagles, and crows can all swoop over fences and find an easy meal. Hawks and eagles will occasionally target young turkeys, and on rare occasions adult turkeys. Crows seldom attack turkeys, but have been known to do so from time to time.

Hawks and eagles tend to be territorial, and not all choose to dine on domestic poultry — as long as they can find other prey. Keep in mind that a hawk or eagle that does not dine on your birds can actually be your best ally in keeping away any such that would.

Habits and clues. These predators hunt during the day. When they attack, they leave a pile of feathers at the strike site. Additionally, they will leave a pile of feathers while feeding on the unfortunate turkey, having no wish to digest the plumes.

Owls

There are many species of owls and they come in a variety of sizes — from very small to quite large. Owls are territorial, and surprisingly, even the small owls will sometimes prey on young turkeys. Some individuals, on the other hand, have no taste for turkey.

Habits and clues. Owls strike at night, flying silently, and may land and walk around seeking prey. They also do not like to eat feathers and so will leave a pile where they strike and where they feed.

Buildings enclosed with wire totally prevent owl attacks, with doors and hatches closed. Roofed structures may have tarps draped over them to prevent a straight line of flight (from outside the pen to the target and then back out again). Aerial obstacles discourage all but the hungriest of owls.

Vultures

Vultures are of little concern, but never say never. Turkey vultures do not attack turkeys, but may be found feeding on dead birds. Black vultures will sometimes attack poults, pecking at their eyes and head, but more often will be seen feeding on dead turkeys.

Prevention

Prevention is the key to avoiding predation and the physical and emotional losses that can occur. First assess which predators are found in your area, and then restrict access by designing fencing that keeps turkeys in and predators out.

Thwarting Flyers

Flying predators can be discouraged. The most important step is to provide the turkeys with some form of cover. This cover may double as a shade source or roosting source. A cover with hanging edges will help limit diving attacks, such as those of the hawk, and horizontal attacks, such as the owl uses.

Other tools to use against flying predators include hanging lengths of string across the yard at random angles. This can be bare string, wire, monofilament fishing line, and even string with bits of streamers that flap in the breeze. An old device used to deter hawks and crows is the hawk chaser. This consists of a 12-inch-long (30 cm) orange-crate slat, or other very thin wood slat, painted red with white angled stripes. A silver circle is painted in the center and the slat is then suspended by 12 inches (30 cm) of fishing line attached to a long pole, which also is set on an angle. The movement of breeze powers this simple device.

I have heard that pigeon fanciers like to keep those globular mirrored balls near their pigeon coops. It is said that diving hawks see their reflections ascending to meet them, and leave rather than risk a fight with this "other hawk."

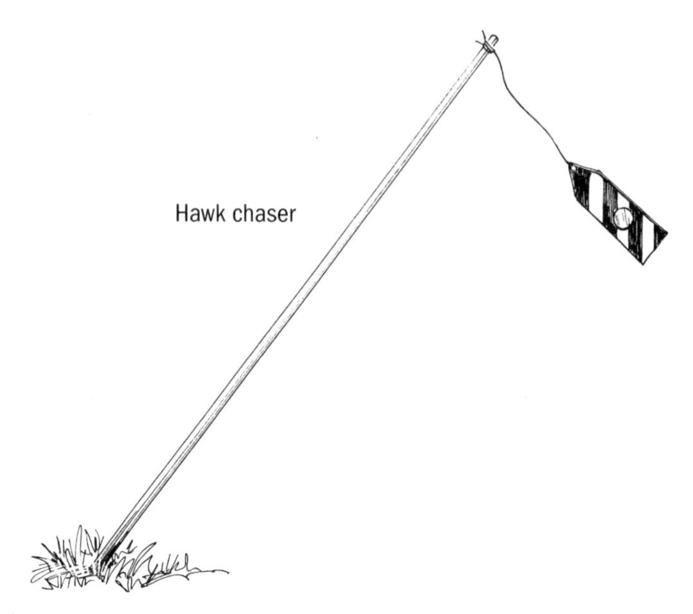

Fencing

Good fencing will resolve 90 percent of your predation threats. It is important for the fence to be designed to keep the turkeys in, as well as to keep predators out. The fence should have good supports to keep it upright and tight, but those supports should not provide entry to walking and climbing predators. The fence should be low to the ground to prevent predators from crawling under it. The fence bottom can be buried to prevent digging, but perhaps the best deterrent is electrified fencing.

Electric Fencing

When it will work for your situation, electric net fencing will keep out most predators while allowing you to rotate the turkeys across your pastures. For turkeys, use the taller versions of this fencing — either 42-inch (106 cm) or 48-inch (122 cm), though the latter is preferred. Mow a strip where you intend to run this type of fencing to ensure that grass and weeds do not touch the fence, and thus grounding out the charge. The charger will be most effective when the soil is moist. When soil is dry, such as during the middle of summer, use more than one grounding rod to increase effectiveness. Moisten the soil near your grounding rod daily when you change your turkeys' water; simply pour the old water over the grounding rod. This will also increase the effectiveness of the fencing.

Electric net fencing should be stretched tight. You can use several bundles to create a fenced pen as large as you need to accommodate your turkeys.

For larger pen sizes, you may find it useful to pound in metal T-posts just outside the corners of the pen, and then attach the corners using nonconductive materials such as baling twine, in order to keep the fence tight. If you use these corner metal T-posts, you may also use them to hold a portable solar charger. In this case, the post holding the charge can act as the grounding rod. A second T-post can be pounded into the ground nearby and connected with a bit of wire to act as a second grounding rod. Electric net fencing works very well, giving you a portable solution to fencing and protecting your turkeys.

Woven Wire

On some farms, permanent fencing may be the practical solution. When you have a choice, woven wire is the best solution for long runs. My personal favorite woven-type wire is one used for horses, with openings 2

FENCING OPTIONS

Double electric. Richard Schock uses strands of electric wire suspended 6 and 12 inches (15 and 30 cm) from the ground around the perimeter of his farm to keep these predators out. This combined with trapping will protect your farm's poultry. Poultry netting works well too.

Angled boards. Irvin Holmes set boards on a 45-degree angle to keep birds from roosting close to the wire; thus predators could not reach in and pull birds to them.

Mesh-wire barricade. Mark Atwood put ¼-inch (0.6 cm) galvanized mesh wire on his ground-level pens — this needs to be up to the height of an adult raccoon. This prevents predators from reaching in and grabbing any young birds or birds that failed to fly up to roost at night.

Remember, you must not only keep the birds in, but you must design pens and brooders so that small predators cannot enter and varmints with dexterous paws cannot reach in.

inches (5 cm) wide by 4 inches (10 cm) high. American wire can be used as well. American wire has tight meshing near the ground and larger openings near the top. In either case, the wire must be stretched tightly and should be very close to the ground—within an inch (2.5 cm) or so. The old saying is that a good fence should be hog tight and goat high—that is, tight and low but high enough to keep in a goat (since goats are able to climb).

Turkey Wire

When you are using a building with permanent outdoor runs for housing your turkeys, another class of fencing will work well. Turkey wire is a form of welded wire that uses 2-inch-wide (5 cm) by 4-inch-high (10 cm) openings. It is usually very thick and can be cut to fit most applications. Welded wire also comes with smaller openings, such as 1 × 2-inch (2.5 × 5 cm), and should be used when small predators are a concern. Chicken wire can be used as well, but chicken wire is very easy for raccoons to unravel, lacks strength to keep out dogs, and will rust easily. Fortunately, chicken wire now comes dipped in a plastic coating. This is a great product because it no longer rusts, at least not for many years, raccoons can no longer unravel it, and it is lightweight and easy to work. If you are using permanent fencing near a building, be sure to extend the fencing

COUNTRY MUSIC

When electric fencing or electrified strands of wire are not an option, you still have a few resources. While studies show that classical music can enhance, very slightly, the growth rate of poultry, seemingly by keeping them calmer, practical applications seem to indicate that raccoons like this form of music. My friend Paul Gilroy of Maryland keeps a radio playing country music near his coops. Paul told me that rock scares away predators but it upsets his poultry. He feels country music works best as it does not upset the poultry, but the raccoons, and other nighttime predators, seem to stay away from his pens because of the sound. Could be that generations of encountering farm boys and girls has instilled the fear of country music into raccoons. Or it could simply be that it sounds most like human activity.

12 inches (30 cm) down into the soil and have pieces that extend 18 inches (46 cm) outward from the pen. This will prevent predators from digging under the fence, as they will dig down to buried wire and be unable to dig under the fence.

Enclosures

When you are using sun porches or a small, freestanding, wire-floored turkey pen raised above the ground, there are a few things you should consider. Keep in mind that some predators will try to reach inside the pen to grab the turkeys. Use 14-gauge, or thicker, galvanized wire mesh with openings of ½ inch (1.2 cm) or less. This is small enough and strong enough to prevent nearly all predators and is easy on the feet of the turkeys.

A turkey at night, on its roost, is at its most vulnerable. Think about where the turkeys are likely to roost, and be sure this spot, or spots, are inaccessible to predators. Boards or cardboard nailed on the end of the roosts at an angle will prevent turkeys from roosting too near walls or windows, and keep them out of reach of predators.

Add Electric

If you are using wire fencing, understand that many critters can climb through the openings or dig under the fencing. Whenever possible, use strands of electric wire to help prevent predators from coming close to your flock. Two strands suspended, one at 6 inches (15 cm) and one at 12 inches (30 cm) from the ground on the outside of the perimeter, will go a long way toward keeping most predators away. Trim weeds away from these strands to prevent them from grounding out and becoming ineffective. Foxes and dogs are usually the biggest predator threat to turkeys (followed closely by raccoons). Electric fences are very helpful at keeping these predators at bay.

Guardian Dogs

Another method works best for me — man's oldest friend. When I got into sheep, more than few years ago, I also ended up adding a sheep guard dog to my farm. This has been a great decision! It turns out that with careful and supervised introductions, sheep guard dogs easily become bird-friendly and willing protectors of poultry. Sheep guard dogs fulfill their role not so much by killing anything, but by their very presence. The dog should be encouraged to walk the perimeter of its property, urinate and defecate along fence lines, and bark at strange things.

My first sheep guard dog was a Maremma, and she once stood all day barking at a hawk — literally for about 5 hours — until the hawk decided the neighboring farm's fields had more tempting fare (rabbits and such). My second dog is half-Akbash, and she is so calm that the wild pheasants view her as no threat. She is a rare example of a dog that bonded with poultry from day one and never tried to play with them.

A number of dog breeds can be useful: Akbash, Anatolian, English Shepherd/Farm Collie, Great Pyrenees, Kangal, Karakachan, Komondor, Kuvasz, Maremma, Sarplaninac, Tatra, and Tibetan Mastiff. My neighbor even had a Labrador retriever that decided he was the protector of all farm critters.

As mentioned in chapter 6, the key to success with a guard dog is to supervise initial contact and to stop or prevent young dogs from playing with the animals under their protection. Guardian dog breeds tend to be independent thinkers, so don't expect them to look to you for direction. In fact, while I have trained most of mine to sit, they rarely sit in a spot

Guard dogs can be very effective, even when they themselves are not yet trustworthy enough to be left unsupervised with their charges. A doubly secure barrier is one with a dog patrolling just outside an electronet fence – predators are unlikely to be able to get past both the dog and the fence.

DON'T ATTRACT PREDATORS

Predator prevention should include reducing or eliminating those things that are likely to attract predators in the first place. Open feed containers must be closed. Wasted feed should be cleaned up nightly if possible. Moving the turkeys around the farm, or daily raking of manure from under roosts and depositing it in a compost heap, will reduce the smell of the turkeys. Any discarded turkey eggs should be cracked open and buried deep in the compost heap. Dead turkeys should be properly disposed — bury, compost, or incinerate immediately. The smell of the turkey manure, their wasted feed, discarded turkey eggs, and any dead turkeys are the four things most likely to attract predators. Your goal is to minimize these attractants as much as feasible.

I indicate but rather one of their choosing — letting me know I may be pack leader, but they are not blind followers.

An excellent guard dog is one that sleeps most of the day. Such a dog is up at night when most predators come around. The smell of the dog, its marking of territory, and its barking are the main reasons predators skirt around the property. All predators know instinctively that if they get hurt they are likely to get sick and die. For this reason, even large predators will usually avoid a farm that has one or more guard dogs.

Other Guardians

Some folks have had great success with noncanine flock protectors. Neutered male llamas and donkeys can make good protectors for your flocks. It takes time to get these vegetarian protectors accustomed to the flock under their care, and you do need to keep these animals out of the turkey feed or they may suffer a health ailment, such as overeating bloat. Since most expensive llamas are guarded by livestock guardian dogs themselves, you can guess what I prefer. But there are options out there that may fit your situation and preference.

In the early part of the 1900s, and for hundreds of years before, farmers often camped out near their turkeys to better protect them. A flock of turkeys had a great economic importance in those days, and prevention of losses took a hands-on approach. This is still a viable option for those willing to camp out.

Your Ecosystem

Having predators is not necessarily a bad thing. A small farm or homestead is an ecosystem. Human endeavor and impact on the land helps shape the ecology of that land — open fields, fencerows, tree stands, and plants available. Many species will strive to live in sync and profit from human landscapes. Predators form an important role in this landscape and help control many wild species — such as rats, mice, rabbits, and deer. Some predators will, through protection of their territory, actually prevent other predators from coming onto your farm to hunt your livestock and poultry. So keep in mind that indiscriminate hunting and trapping are more likely to create problems than solve them.

Proactive Protection

There are a few important points to keep in mind when considering hunting and trapping predators. First, know the law. Some animals are protected by state and federal laws and either cannot be killed or require

THINK LIKE A PREDATOR

Another important step in preventing predation is to walk your property. As you walk along, notice any trails on the ground, especially those leading through a fencerow. Notice any tracks you see and try to take a picture (with something, like a coin, for scale) for later identification. Note any burrows you find, especially those with fresh soil near their entrance.

Think like a predator — how could you sneak, unobserved, to your turkey pens? Are there hedgerows, overgrown fencerows, or brush that would make excellent cover? Notice any draws and the location of water, creeks, and streams. Many predators will arrive on your farm by walking along waterways — especially raccoons, mink, and weasels. Know where the predators will come from and how they are likely to arrive.

special permits. Second, if you opt to trap or poison, know that doing so not only poses a risk to any of your or your neighbor's animals that come into contact with such, but may be illegal in your area. Again, know the law. If you choose to live trap an animal, such as a raccoon, be sure to drive more than 10 miles away when you release and try not to create a problem for the farmer near the drop-off site. I had one friend that caught the same raccoon several times — more miles is a good idea.

For those who can and do legally shoot a predator, consider what you can and should do with the carcass. There may be local laws on proper disposal of predators. When there are not, let me clue you in on another very old, useful method of deterring predators. I have a friend who places parts of the dead raccoons around his coops. He has woods very near his birds, and the smell of the deceased raccoons seems to send a message to others of its kind that this is not a safe area to hunt. While a bit gruesome, it seems to keep the predators away for great lengths of time. I would also say that you should ask yourself these two questions: What made your birds vulnerable to predation in the first place? How many birds can you afford to lose before you feel you need to use a "permanent" solution?

Turkey Nature

When trying to protect turkeys, remember the nature of the birds. Turkeys flock together; roost at night; prefer to have available access to cover; move through brush by finding a way with their heads and then using their wedge shapes to force their way; and, when spooked at night, will come off the roost and possibly stampede in the dark. Startled turkeys will seek to return to the flock and to the roost area — be sure to limit the possibility of startled turkeys landing on the outside of their fence.

Frightened turkeys can and do stampede. It is a part of a turkey's nature to make a sudden burst of flight from a dangerous location. This burst will be relatively short and is often followed by quiet and stillness. Essentially, the turkey's instincts tell it to move quickly away from the danger, and then avoid attracting more interest by slowing down and seeking a secluded hiding spot. A handful of turkeys stampeding is a small problem, but when you have 100 or more birds stampeding, they may well breach fences and cause harm to each other.

When housed turkeys are startled, they will fly against walls, the roof, and roosts — causing bruises and even broken legs and wings. The startled turkeys often will pile up in corners, especially in the dark, and many can be lost to smothering. Stampeding can result from such simple stimuli as a strange noise, sudden bright lights, sudden movement, or prowling predators.

When you realize the nature of turkeys, you can better understand how to protect your turkeys. Outdoor roost placement should be near the center of the pen to prevent startled turkeys from landing on the wrong side of the fence and becoming easy dinner as they try to force their way back through the fence, head first. Fencing should be tight and closely woven, and the area around the roost should be free of obstacles that could harm a turkey trying to land in the darkness. A simple solar-powered light, or several, can be placed near the turkeys' roost and will go a long way toward preventing fright and pileups. Lights also help turkeys find their way back to the roosts and discourage nighttime predators.

◆ ◆ ◆

You can prevent predation if you plan ahead and understand the challenges you face.

- Plan pens to keep predators out and turkeys in.
- Reduce those things, like wasted feed or dead birds, that attract predators.
- Know your farm's terrain and realize how and where a predator might breach your security.
- Enlist a guard, such as a trusty dog or donkey, or electrify the perimeter to help predators choose not to risk a turkey meal.
- Be sure to have a reason to hunt or trap — don't mistakenly rid your farm of the raccoon or hawk that has no taste for turkey.
- And last, if you do trap, be sure to make the most of this effort — don't drop the displaced critter on your neighbor's doorstep or close enough that it will return.

With a little planning, you can live in harmony with wild critters and have a successful turkey enterprise as well.

9

Killing and Processing

TURKEYS SHOULD BE FINISHED and ready for processing at twenty-four weeks of age for old varieties and eighteen weeks of age for newer, heavier varieties. Hens are usually processed at younger ages than toms. Turkey broilers or fryer-roasters are usually animals of the same strain used to produce heavier carcasses but are processed at younger ages, such as eight to ten weeks of age. The precise age for finishing and processing depends on the turkey variety and strain, the feeding program, and other factors.

Assessing Readiness for Processing

To assess whether a bird is in prime condition and ready to be processed, see if it is free of pinfeathers. The bird is "ready" when the feathers are easy to remove. Pinfeathers are immature feathers that do not protrude or may have just pierced the skin. Short protruding feathers have the appearance of a quill with no plume. They are unattractive, particularly in varieties with dark feathers, and cause downgrading when present in finished market birds. If the bird is not going to be marketed but, rather, consumed at home, the pinfeathers may stay in place; however, if the presence of pinfeathers is considered to be a drawback, it is best to delay dressing those birds until the feathering improves.

Evaluating Degree of Fat Covering

You must also check the degree of fat covering.

1. Pull a few feathers from the thinly feathered area of the breast, at a point about halfway between the front end of the breastbone and the base of the wing.

2. Take a fold of skin between the thumb and forefinger of each hand.

3. Examine for thickness and coloration. On a prime turkey, the skin fold is white or yellowish white and quite thick. Well-fattened birds have thick, cream-colored skin, while underfattened birds have thin (often paper-thin) skin that is semitransparent and tends to be reddish.

Care Before Killing

Careless handling can cause birds to pile up and trample each other, resulting in injuries. Recent injuries may appear red at the bruise site; old injuries are bluish green. Such defects detract from the dressed appearance. Always catch birds properly.

Withhold feed from the birds for approximately 10 hours before killing; however, do not withdraw water or excessive dehydration may occur. Removing the feed enables the crop and intestines to empty before killing and makes the job of eviscerating much cleaner and easier. Remove the birds to be feed-restricted from the pen, and put them into coops containing wire or slat bottoms to keep them away from feed, litter, feathers, and manure. After catching the birds, keep them in a comfortable, well-ventilated place prior to killing. Overheating or lack of oxygen can cause poor bleeding and result in bluish, discolored carcasses.

PROPER BIRD-CATCHING TECHNIQUE

1. Grab the legs between the feet and hock joints with one hand.
2. Straighten the legs to lock the hock joints. Don't grasp the legs at the feathered area above the hock joints, as this may cause skin discoloration.
3. After catching the bird by the shanks, hold one wing at the base with the other hand. This immobilizes the bird effectively. It also gives the handler control of the bird and prevents injuries and bruising.

GETTING STARTED

Plan to process the turkey in a clean, well-lit area that has a water supply and no flies. It is helpful to have flat surfaces that can be easily cleaned, and suitable containers for handling the offal (or waste by-products).

Processing Area and Equipment

Home processing of just a few birds requires little in the way of special facilities or equipment; but if a fairly large number of turkeys are to be dressed, you should have an adequate area and some special equipment, such as a mechanical picker.

Process your poultry in as sanitary a manner as possible. It is important to prevent contamination of the carcasses. One of the most common sources of contamination is the contents of the intestine, although dirty facilities, equipment, and people can also contribute. Contamination reduces quality and shelf life — that is, the period before spoilage begins.

At best, the processing job is a messy one. Ideally, there should be two rooms available for processing. If several birds are to be done at one time, use one room for killing and plucking the birds and the other for finishing, eviscerating, and packaging. If this is not possible, or just a small number of birds is involved, do the killing and plucking in one operation, clean the room, and then draw and package the birds as a second operation. When you can use only one room, following this method makes the procedure far more sanitary. Good organization makes the process go more smoothly.

Ideally, the processing equipment should be made of metal or other impervious material to facilitate plant cleaning and sanitation. The processing plant requires a plentiful water supply, at least 5.5 to 10 gallons (21–38 L) of water per dressed turkey.

Killing Cones or Shackles

If only a few birds are to be dressed, a shackle for hanging can be made from a strong cord with a block of wood, 1 × 2 inches (2.55 × 5.1 cm) square, attached to the lower end. A half hitch is made around both legs and the bird is suspended upside down. The block will prevent the cord from pulling through.

Commercial and semicommercial dressing plants use metal shackles that hold the legs apart and allow easy plucking. Some producers make their own shackles out of heavy-gauge wire.

Others prefer to use killing cones, which are similar to funnels. The bird is put into the cone with its head protruding through the lower end. This restrains the bird and reduces struggling, which can lead to bruising or broken bones.

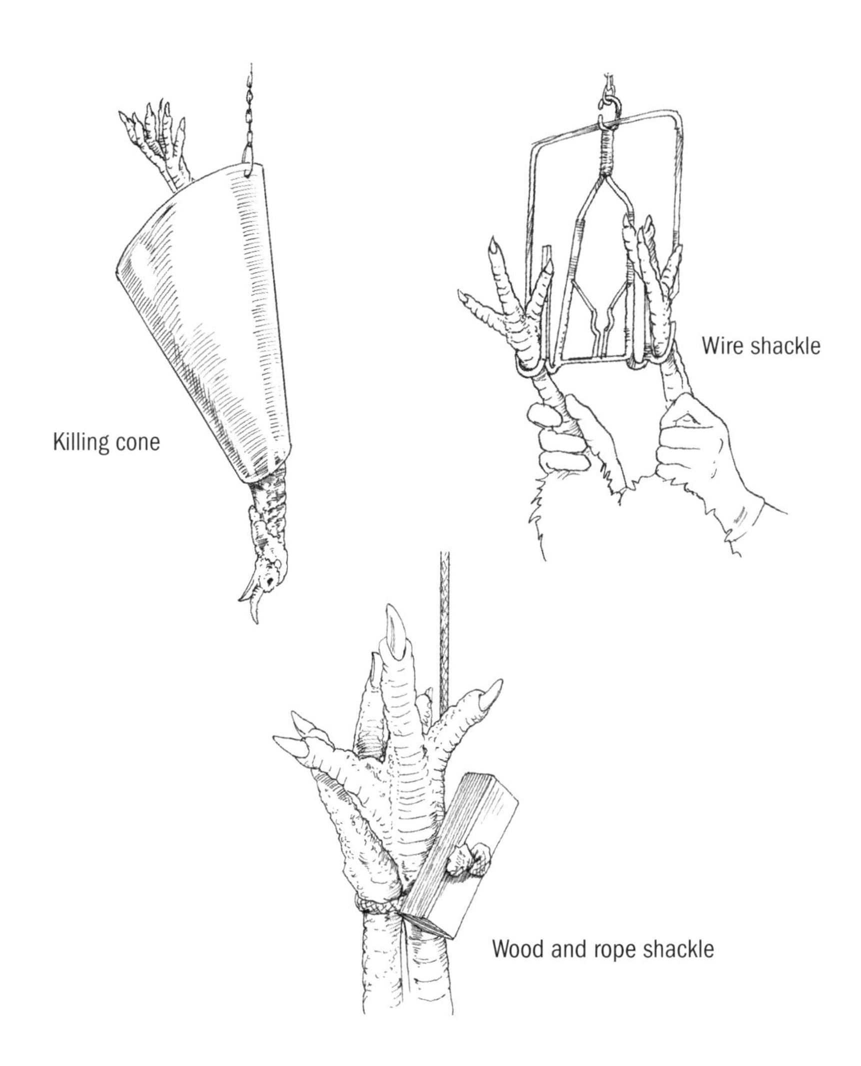

Weights

A weighted blood cup or a simple weight attached to the lower beak of the bird prevents it from struggling and splashing blood. You can make the device from a window weight attached with a sharp hook to the lower beak. The blood cup is not used when killing cones are available.

You can make a blood cup from a 2-quart (1.9 L) can. Solder a sharp-pointed, heavy wire to the can. The wire hooks through the lower beak. Weight the cup with concrete or heavy stones.

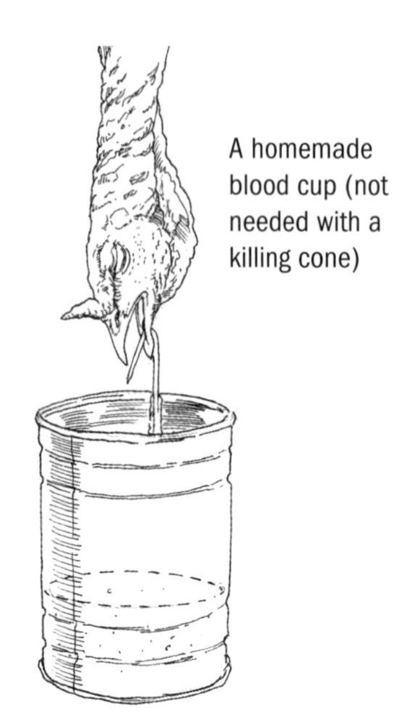

A homemade blood cup (not needed with a killing cone)

Knives

Just about any type of knife can be used for dressing poultry. There are special knives for killing, boning, and pinning. Six-inch (15 cm) boning knives work well. If the birds are to be brained, then use a thin sticking or killing knife. Make sure all knives, especially the killing knives, are very sharp.

COMMON POULTRY KNIVES AND IMPLEMENTS

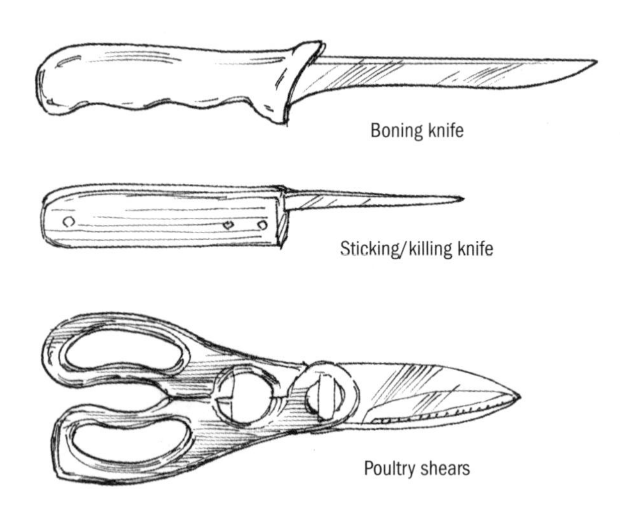

Scalding Tank

When the birds are to be scalded and only a few birds are to be dressed, a 10- to 20-gallon (37.9–75.7 L) garbage can, or any other clean container of suitable size, is satisfactory. When a considerable amount of dressing is being done, a thermostatically controlled scalding vat is preferred. In the absence of the automatically controlled vat, hot water can be continually heated and the vat replenished as required to maintain a desired temperature.

Homemade, electrically heated scalding vat with thermostatic control

Thermometer

Accurate temperatures are important for scalding. Acquire a good, rugged dairy thermometer, a candy thermometer, or a floating thermometer that accurately registers temperatures between 120° and 150°F (48.9° and 65.6°C).

Killing

There are many methods of killing turkeys. The first method that follows is perhaps the simplest. An alternate, slightly more difficult method is also provided. Both involve severing the bird's jugular vein. The jugular vein needs to be thoroughly severed to ensure that the birds are well bled. With either method, make sure the killing knife is razor sharp, which will allow for a more humane kill. A bird that is not well bled will have a purplish skin color that seriously affects the bird's dressed appearance and marketability.

Method 1

1. Suspend the turkey by its feet with a rope or metal shackle, or place it in a killing cone.
2. Hold the head with one hand and pull it down to exert slight tension, which steadies the bird.
3. With a sharp knife, sever the jugular vein just behind the mandibles. This can be done by inserting the knife into the neck close to the neck bone, turning the knife outward, and severing the jugular. It may also be done by cutting from the outside.

Method 2

The jugular vein can also be severed from inside the mouth; this is slightly more difficult than the previous method.

1. Hold the head in one hand, with your fingers grasping the sides of the neck, taking care not to squeeze the jugular vein.
2. Make a strong, deep cut across the throat from the outside close to the head so that both branches of the jugular vein are severed cleanly at or close to the junction. *Warning:* Be sure to hold the head so your fingers do not gct in thc way.

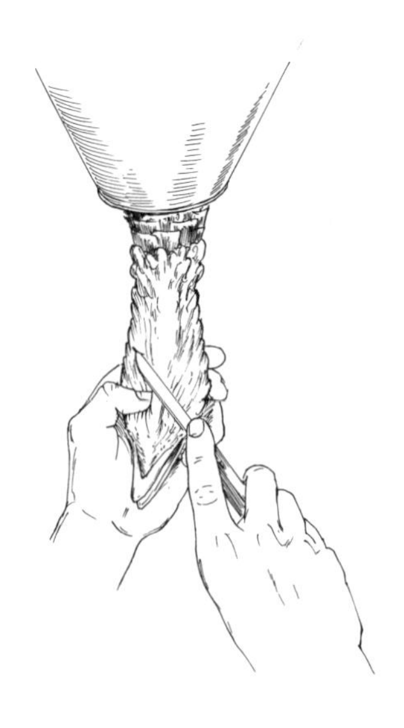

3. Do not grasp the wings or legs to the extent that you restrict blood flow from these parts. Incomplete bleeding results in a poor-appearing carcass.

Birds can be slaughtered either conscious or unconscious. Combination stunning and killing knives are frequently used. The knife has an electrical component with a button. The knife is held next to the bird's head, and the bird is stunned when the button or switch is on. The stunning renders the bird unconscious. The switch is turned off, and the bird is slaughtered.

Debraining

This process loosens the feathers so that it is easier to pluck the birds. Debraining is done before the jugular vein is cut in birds that are to be dry-picked, but it may also be done when the carcasses are to be scalded (see below) or to make feather removal even easier. Though dry picking is slower, the outer layer of skin is not removed, making for a fine-appearing dressed carcass.

How to Debrain

This procedure is tricky and requires considerable practice before proficiency is achieved.

1. Insert the knife through the groove or cleft in the roof of the mouth.
2. Push the knife through to the rear of the skull so that it pierces the rear lobe of the brain as shown.
3. Rotate the knife in a one-quarter turn. This kills the bird and loosens the feathers.

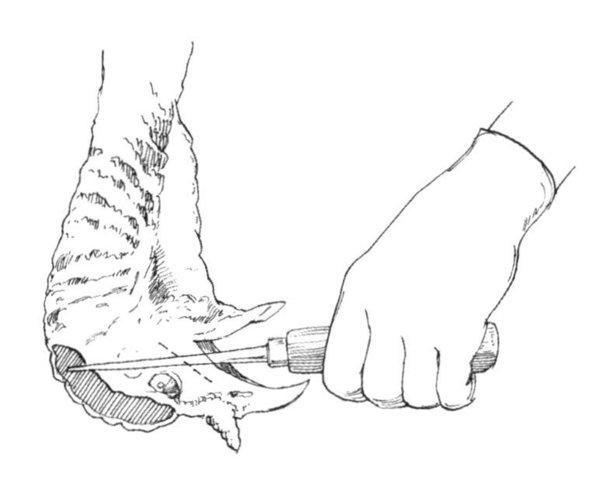

A characteristic squawk and shudder indicate an effective stick.

Scalding

There are two methods of scalding: subscalding and semiscalding. Both work equally well.

Subscalding

As soon as the bird is dead and bleeding is complete (usually 2 to 3 minutes), loosen the feathers using the subscald method. Dunk the bird in water at approximately 140°F (60.9°C) for about 30 seconds. The subscald method makes it easy to remove the feathers and gives the skin a uniform color. The skin surface tends to be moist and sticky and will discolor if not kept wet and covered. For the scald to be effective, slosh the bird up and down in the water to get the water around the follicles at the base of the feathers.

WATER TEMPERATURE AND TIMING ARE CRITICAL

For an effective semiscald, the water temperature must be maintained within the narrow range of 125° to 130°F (51.7°– 54.4°C). Ideal temperature is 126°F (52°C) for smaller turkeys and 128°F (53°C) for larger ones.

Scald turkeys for 30 seconds each. If the water is a little cool or the scalding time too short, the feathers will not loosen enough for easy picking. If the feathers are difficult to pull out, skin tears can result. If the water is too hot or the scalding time is too long, the bird will have an overscalded or patchy appearance.

To make picking easier and faster, cover the turkeys with dry sawdust for a few seconds to dry the feathers before plucking.

Turkeys should be chilled immediately after butchering. Either cool air or ice water work well to bring down the temperature. If turkeys are held fresh too long, oxidation of the fat will occur and they will develop a fishy flavor and odor in the kidneys and drumsticks.

Withhold fish oils and meal from turkeys for 2 to 3 months before slaughter, to avoid off-flavor and smell.

Semiscalding

Another method that is sometimes used is semiscalding. The bird is scalded for 30 to 60 seconds in water 125° to 130°F (51.7°– 54.4°C). With the semiscald method, the feathers loosen but the temperature is not hot enough to destroy the outside layer, or skin cuticle. Thus, the carcasses look more like dry-picked birds.

Plucking

If available, a rubber-fingered plucking machine can remove the feathers as well as the **cuticle** (or bloom), which is the thin, outer layer of the skin. Remaining pinfeathers are removed by hand. Don't let the skin dry out or it will become discolored. If they are not immediately eviscerated, put the birds in cold running water.

Hand-Plucking

Hand-pluck feathers in this way:

1. Rehang the bird on the shackle.
2. With a twisting motion, remove the large wing and tail feathers first.
3. Remove the remainder of the feathers as quickly as possible in small bunches to avoid tearing the skin.

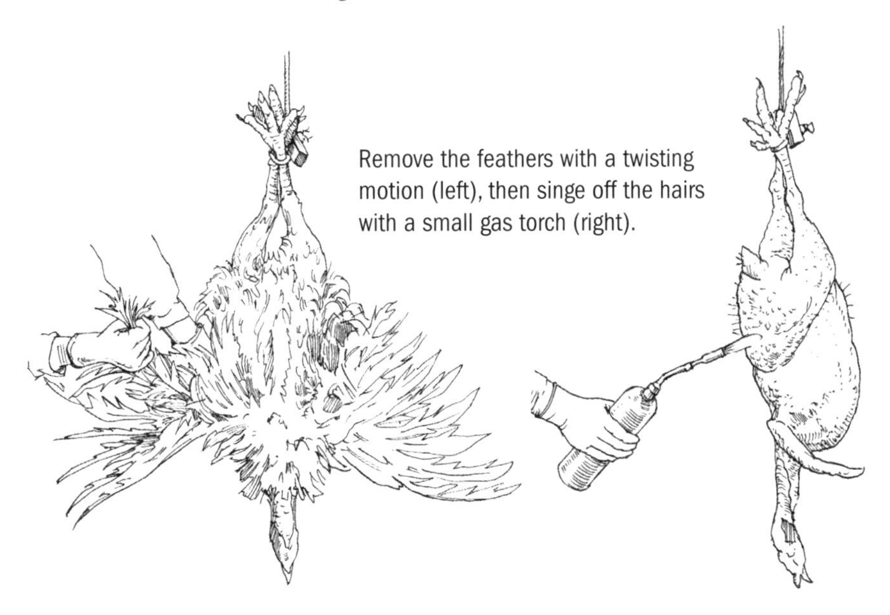

Remove the feathers with a twisting motion (left), then singe off the hairs with a small gas torch (right).

Pinning and Singeing

Pinfeathers, the tiny, immature feathers, are best removed under a slow stream of cold tap water. Use slight pressure and a rubbing motion. You can use a pinning knife or a dull knife to pluck the feathers that are difficult to remove. By applying pressure between the knife and the thumb, you can squeeze out the pinfeathers. The most difficult may have to be pulled. Usually, turkeys have a few hairlike feathers left after they have been hand-plucked. You can singe these hairs with an open flame. A small gas torch works well. Do not apply the flame directly on the carcass to avoid scorching the skin.

The pinning and singeing process may sound time consuming, but it will improve the appearance of the carcass as well as increase customer demand for your product.

Eviscerating

After picking and singeing, wash the carcasses in clean, cool water. The carcasses are ready for evisceration as soon as they are washed. Some prefer to cool the poultry first because, after cooling, eviscerating is somewhat easier and cleaner. Others eviscerate and then place the birds in ice water or cool water that is constantly replenished. There are many methods of eviscerating poultry, but the most important part of the process is to keep your working area and equipment clean.

The parts of the turkey are removed in the following order: (1) tendons (optional); (2) shanks and feet; (3) oil sac (preen gland); (4) crop, windpipe, gullet, and neck; (5) lungs, liver, and heart attachments; (6) gonads and kidneys.

MATERIALS

Only a few items are needed for evisceration:

- Sharp, stiff-bladed boning knife
- Hook (if the leg tendons are to be pulled)
- Solid block or bench on which to work
- A piece of heavy parchment paper or meat paper that is laid on the working surface and changed as necessary

WEAR GLOVES

Practice good sanitation when handling and processing the carcass — wear disposable surgical, latex, or vinyl gloves.

Removing the Tendons, Shanks and Feet, and Oil Sac

Sometimes the tendons are removed from the drumsticks before removing the shanks and feet. Removal of the tendons makes carving and eating the drumsticks easier. By cutting the skin along the shank, the tendons that extend through the back of the leg may be exposed and twisted out with a hook or a special tendon puller, if available.

Remove the oil sac on the back near the tail, as it sometimes gives the meat a peculiar flavor. This is removed with a wedge-shaped cut.

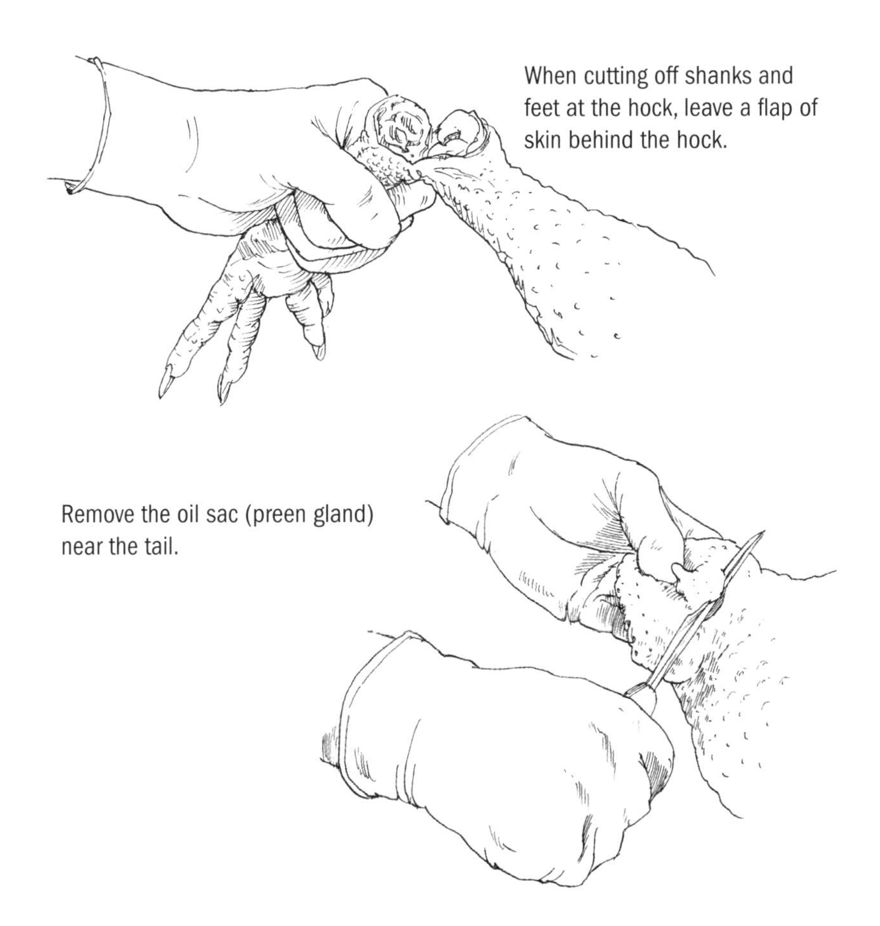

When cutting off shanks and feet at the hock, leave a flap of skin behind the hock.

Remove the oil sac (preen gland) near the tail.

Removing the Crop, Windpipe, Gullet, and Neck

1. To remove the crop, first cut off the head.

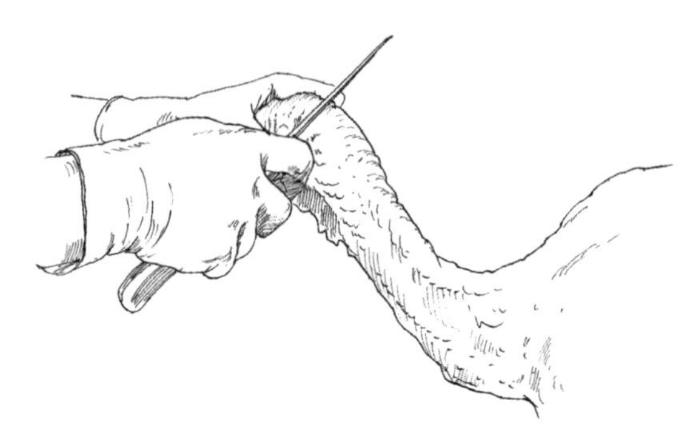

2. Slit the skin down the back of the neck to a point between the wings.

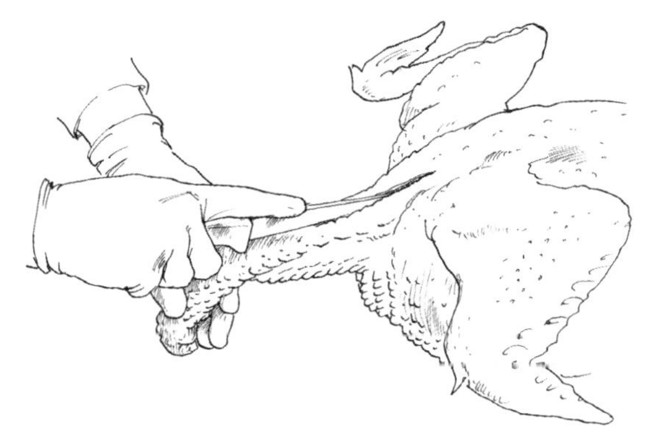

3. Separate the skin from the neck and then from the gullet and windpipe.
4. Follow the gullet to the crop and remove, being careful to cut below it.

5. Cut off the neck as close to the shoulders as possible. A pair of heavy shears is handy for this purpose; you can also sever the neck by cutting around its base with a knife and then breaking and removing with a twisting motion.

6. Loosen the vent by making a circular cut around it. Do this carefully to avoid cutting into the intestines.

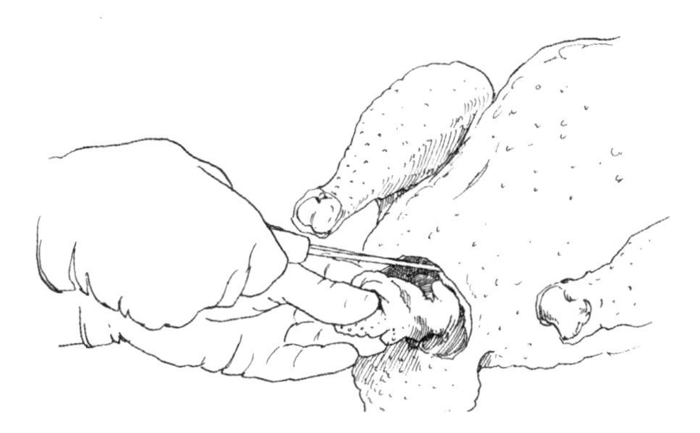

Removing Abdominal Contents

1. Make a short, horizontal cut 1½ to 2 inches (3.8–5.1 cm) between the vent and the tip of the keel bone; make the horizontal cut about 3 inches (7.6 cm) long.

2. Break the lungs, liver, and heart attachments carefully by inserting the hand through the rear opening.
3. Loosen the intestines by working the fingers around them and breaking the tissues that hold them.
4. Remove the viscera through the rear opening in one mass by hooking two fingers over the gizzard, cupping thc hand, and using a gentle pulling and slight twisting motion.
5. Remove the lungs, gonads, and kidneys. The lungs are attached to the ribs on either side of the backbone. These can be removed by using the index finger to break the tissues attaching them to the ribs. Insert a finger between the ribs and scrape the lungs loose. The lungs appear pink and spongy. The gonads are also attached to the backbone.

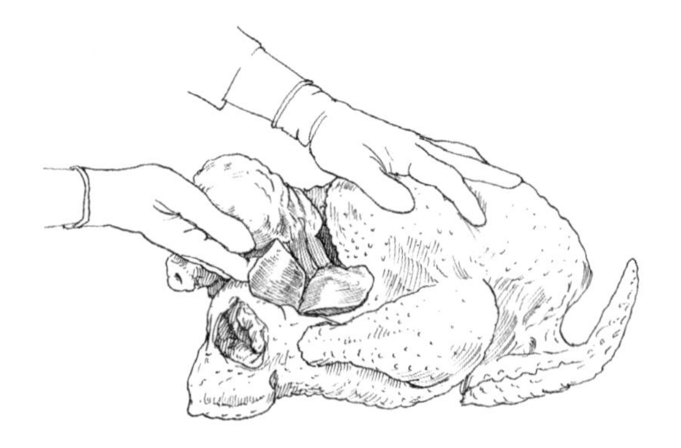

Washing the Carcass

1. Wash the inside of the carcass using water from a faucet or from a clean hose.
2. Wash the outside to remove any adhering dirt, loose skin, pinfeathers, blood, or singed hairs.
3. Hang the bird to drain the water from the body cavity.

Cleaning the Giblets

Remove the gallbladder, which is the green sac attached to the liver, without breaking it. If the gallbladder breaks during removal of the viscera or while cleaning the liver, the bile is likely to give a bitter, unpleasant taste to any part it contacts and will cause a green discoloration.

If you are careful, a cool gizzard can be cleaned without breaking the inner lining. Cut carefully through the thick muscle until a light streak is observed. Do not cut into the inner sac or the gizzard lining. The gizzard muscle may then be pulled apart with the thumbs, and the sac and its contents will be removed unbroken, if you're lucky.

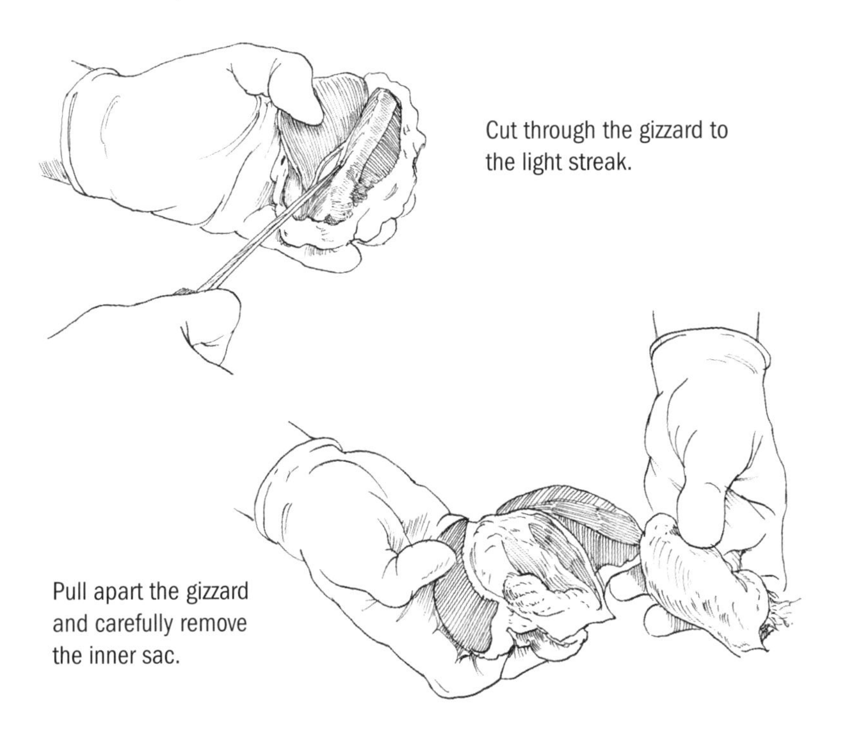
Cut through the gizzard to the light streak.

Pull apart the gizzard and carefully remove the inner sac.

Trussing

A properly trussed bird appears neat when packaged. Proper trussing also conserves juices and flavors during roasting. The simplest method to truss a bird follows:

1. Tuck the hock joints under the strip of skin between the vent opening and the cut from which the viscera were removed.

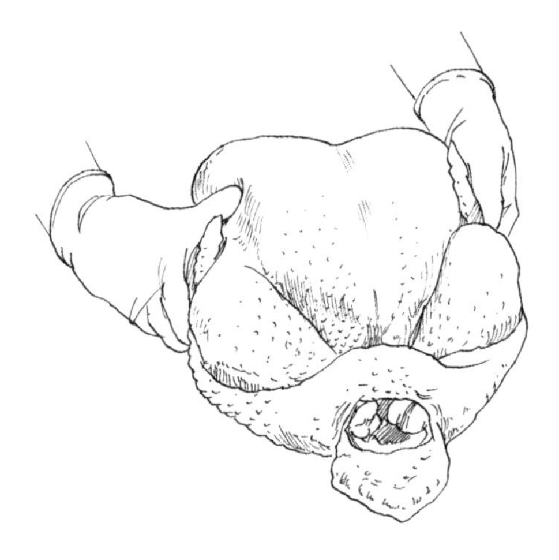

2. Turkeys are usually packaged with the wings in a natural position. Another option is to draw back the neck flap between the shoulders and fold the wing tips over the shoulders to hold the skin in place.

Chilling and Packaging

The next phase of processing your turkeys involves chilling the birds in water, wrapping the giblets, and wrapping the turkey.

Chilling the Birds

Cool the birds as soon as possible after killing. If cooling is done slowly, bacteria can develop and cause spoilage and undesirable flavors.

You can cool poultry with water if air-cooling is not possible. If dressed with excessively high, scalding temperatures or for too long a period, the skin of air-cooled birds may be blotchy and discolored. When scalding temperatures are too high, water-cooling is the preferred method of cooling the carcasses. Dressed birds may be cooled in tanks of ice water or

GUIDELINES FOR AIR-COOLING

If birds are to be air-cooled, the air temperature should be from 30 to 35°F (–1°–1.7°C). The time required to cool the carcasses depends on the size of the bird and the temperature of the air. Birds to be air-cooled should always be packaged to avoid discoloration.

under cold running water. The important factor is to maintain a constant temperature of 34 to 40°F (1°–4.4°C). For a bird's internal temperature to become that cool, it must remain in the water for 5 to 10 hours, again depending on the size of the carcass. If eaten or frozen immediately after dressing, carcasses tend to be tougher than if aged for a while.

Remove the carcasses from the water and hang them up to dry for 10 to 30 minutes before packaging. Make every effort to remove all of the water from the body cavity before putting a bird in the bag.

Wrapping the Giblets

Spoiled giblets can spoil the entire carcass, so always wrap them well. Use this method to ensure proper packaging:

1. Wrap the giblets — that is, the neck, gizzard, heart, and liver — in a sheet of wax paper or a small plastic bag.
2. Stuff the giblets into the body cavity or under the neck skin.

Bags for Wrapping Poultry

There are two types of bags available for poultry. One is the so-called Cryovac bag (W. R. Grace and Co., Duncan, South Carolina); the other is a common plastic bag. When the turkey is placed in the Cryovac bag and then boiled, the bag shrinks and adheres to the bird. Not only do these bags make a nice-appearing package, but they also help to reduce the amount of water that builds up during the freezing process.

Good-quality plastic bags are also available and do a satisfactory job of maintaining quality in frozen, dressed poultry. The bags should be impermeable to moisture to prevent dehydration during freezing, which causes toughness.

After chilling, insert the carcass into a plastic bag. Suck out the air with a vacuum cleaner or a plastic hose (left) and secure the bag with a twist tie (right).

Birds to be bagged should be trussed thoroughly (see above), then inserted front-end first into the plastic bag. After the bird is in the bag, you can remove excess air by using a vacuum cleaner or by inserting a flexible hose into the top of the bag and then creating a vacuum. Merely keep the bag snug around the hose or vacuum cleaner, then suck the air out of the bag. Twist the bag several times and secure it with a twist tie or a rubber band.

DRESSING PERCENTAGE

Live to eviscerated weight*

Type of bird	Live weight in lbs	Percentage of blood and feathers	Eviscerated percentage lost
Broilers and fryers	5–6	7	25
Hens — small	10–12	6	18
Hens — medium	12–15	6	20
Hens — large	15–20	7	20
Toms — small	20–25	7	20
Toms — medium	25–30	7.5	20
Toms — large	35–40	7.5	19

*1 lb = 0.454 kg.

Source: Estimated using method reported by J. Brake et al. "Relationship of sex, strain, and body weight to carcass yield and offal production in turkeys." *Poultry Science* 74.1 (1995):161–68.

Fresh-dressed, ready-to-cook turkeys have a shelf life of approximately 10 days if refrigerated at a temperature of 29 to 34°F (–1.7–1°C). If you plan to freeze your turkey, do it by the third day after it is dressed and chilled. Chill the poultry to below 40F (4.4°C) before placing it in the freezer.

The amount of weight a turkey loses between slaughter and dressing varies by age and the type.

State and Federal Grading and Inspection

Some processors of poultry that is sold off the farm are subject to the Poultry Products Inspection Act. There are exemptions for small producers, and regulations may vary among states. Check the regulations that apply in your area. For information on grading and inspection programs and how they affect you, contact your state department of agriculture.

◆ ◆ ◆

COLOR IN THE VANES

A pigment called melanin gives feathers color and can stain the skin of the turkey when being plucked. Melanin also makes the vanes, or flat webbed part of feathers, strong and durable.

10

Breeding

THE REPRODUCTION OF TURKEYS must be properly managed so they may be useful and profitable, and so that generations of the population will remain viable and healthy. Reproduction of the next generation is the only way in which a species, a breed, or a variety can ensure its continued existence. Domestic turkeys rely on our management to sort out the details of culling and the selection of breeding stock as well as the management of population genetics over generations. It is this interaction that forms the basis of animal agriculture.

Nature's Way

In the wild, turkeys are seasonal breeders that have a mating season running from about February through May. Just as for other poultry and seasonal mammals, increasing sunlight triggers increased hormone production and the onset of breeding season. The natural breeding season encompasses a period of time such that the vitamin-rich foodstuffs of spring are available just in time to be incorporated into the nutrients of the eggs, and later to nourish the offspring. It is no accident that rich food is available in great quantities during breeding and growing, and that the animals reach mature size as weather turns cold and food becomes less plentiful. One should note that there is a relationship between reproduction, the environment, and the seasons.

As mating season begins, the males attract females through a combination of gobbling, "drumming," displaying, and strutting. Males begin gobbling early in the morning while still on the roost. About 15 minutes before sunrise, the males will drop off the roost and begin displaying and gobbling in earnest.

As the males display, they grow increasingly excited. Frequent sharp expulsions of air — or sneezes — precede ruffling of feathers. The tom's face, throat, head, and snood may range among red, blue, and white—white representing a male that is very excited. Males will eye each other up and use a combination of stares and strutting while displaying to intimidate one another.

Social Dynamics

Only dominant males are free to mate. While the dominant male is strutting and drumming, the sibling males stand and display — much like backup singers. This cooperative courtship increases the number of poults produced by the dominant male. If the sibling male is an actual brother, and not simply a fellow turkey from another hen hatched and raised with the dominant male, then this works to his advantage. His brother shares at least half his DNA, and a larger percentage of offspring will have this DNA than if the male did not participate in cooperative courtship.

Squatting behavior is a clear sign of a hen's readiness to mate.

Toms can be extremely aggressive during breeding season. It is not unusual for two toms to wound each other on their heads, throats, wattles, and snoods. A high-pitched trill often precedes aggression — sparring ensues and the toms will peck each other with their beaks, flog each other with their spurs, and rake each other with their talons. In some cases, a tom may lose an eye. If a sibling tom is present, he will help to subdue the competitor and drive him away. Upon victory, the winning tom usually begins to strut and display.

The hens will come and watch the toms strut and display. All the criteria weighing in a hen's choice of mate are not completely clear, but they do seem to be drawn toward males that appear large and are dominant.

Mating, Nesting, and Incubating

Once a hen has selected a mate, she will come and crouch before him, lowering her tail. The tom will then mount the hen, forcing her to the ground. As he lowers his tail, she raises hers and moves it to one side, they touch cloaca to cloaca, and he injects his sperm. Although the sperm will remain viable for up to 8 weeks, hens will mate several times — largely with the same tom. Toms will mate with many hens.

After mating, the hen turkey seeks out a good nesting site. Turkeys are ground nesters and prefer open woods with some brush. The nest is usually secluded from the other turkeys and protected from behind,

COURTSHIP TACTICS

Male turkeys will inhale and spread their feathers to appear larger, strut, gobble, and periodically drum.

A male's **gobble** will carry as far as a mile; females will often travel to see if the male is a worthy suitor.

Drumming is a low-pitched sound caused by the male moving air in and out of the air sac on his chest.

Males **display** by dropping their wings, tips to the ground, raising and fanning their tails, and raising their body feathers.

Strutting describes the male shuffling his feet while displaying, and often includes some drumming and ruffling of feathers.

which allows the hen to observe all activity within her 300-degree field of vision with no fear of being attacked from the rear.

The hen will lay 10 to 16 eggs, about one per day, before attempting to incubate. Once she has accumulated enough eggs, and if she still feels secure, she will begin the process of incubation. If everything goes well, she will emerge 28 days later with a brood of poults.

The hen turkey is very vulnerable during incubation. She must rely on stillness, the camouflage of her plumage with her surroundings, and a good choice of nesting location in order to survive and successfully produce a brood of poults. While toms take no direct role in protecting the hens during this period, and in fact do not even help raise the brood, the continuation of their courtship rituals helps draw the attention of daytime predators to themselves and away from the hens.

Human Impact

Wild turkeys must fend for themselves, and the details of population control and selection are managed by nature. Domestic turkeys are the result of interaction with mankind. Rather than nature making culling decisions, offering foodstuffs and shelter, or providing mates, producers make these sorts of decisions for their domestic turkeys and in doing so, shape the turkey population.

When we manage turkeys, we play a part in a symbiotic relationship that benefits both the turkeys and ourselves. The turkeys benefit by being able to reproduce in safety, often in large numbers, having food provided or made available to them, and being able to continue to survive for many generations. We benefit by having a reliable food source, by having workers that will happily reduce insect populations and help us manage our land, and through the fertility of the turkey manure — spread with little labor on our part when the turkeys are kept on pasture.

In the wild, only individuals that succeed in meeting their own nutritional needs survive. Of those that survive, only the dominant males are able to reproduce — and those males having siblings that cooperate reproduce at a higher level. Only clever females successfully incubate and brood a clutch of poults. Predator avoidance is a skill needed by wild turkeys of all ages. Natural selection in wild turkeys favors those individuals that can find their own food, avoid predators, and reproduce successfully. We can say this requires turkeys with awareness, some intelligence, and robust health, and we can say natural selection favors cooperation among sibling toms and, of course, natural mating and brooding.

Selection controlled by humans favors individuals that meet our purposes best. Since turkeys are primarily utilized for food, this means a larger, meatier body is preferred. In fact, there are technically two distinct body types that do meet and have met our needs — the heritage and the industrial type — the first selected for outdoor production and the second for indoor production.

Heritage turkeys have long legs and a somewhat upright carriage. They have been selected with more robust fleshing than their wild counterparts — possible only if humans help the turkeys find enough calories to support their growth. Heritage turkeys are designed for natural mating and ranging behavior.

Industrial turkeys have short legs, a short breastbone, and horizontal body carriage. They have been selected to produce incredible amounts of flesh—possible only when enough calories are totally supplied. Industrial turkeys are designed such that they cannot physically mate and they range only small distances.

Industrial type (left) compared with wild type (right).

Human choices shape both the bodies and the abilities of turkeys. Understand that care and selection choices have an impact on which individuals survive and reproduce. The better we understand that selection for large size reduces other attributes — such as ranging ability, survival instincts, or traits such as disease resistance — the better managers of poultry we will become.

The Purpose of Breeding

There are three goals of breeding:

- The production of individuals for a specific purpose
- Improvement of the population to suit its purpose better
- Continuation of the population

The first is easy enough to understand — if you need 1,000 turkeys to grow and then process for the Thanksgiving market, someone must produce the poults. There is little to add to this point, except that your choice of source will have a lot to do with the overall quality of the end product. The other two purposes are worthy of more discussion.

Improving Quality

Breeding to improve a population is the most commonly associated purpose of breeding. People advertise breeding stock using slogans such as "Best"; "You have to get some of these genetics"; "Improved Strain"; and "Faster gains." What they are trying to communicate is the desirability of their stock so that they may sell more animals, but what they are inferring is that their population is "better." The basis for improvement favors a small number of individuals and thus a reduction in genetic material.

This form of breeding can be summed up as "Breed from the best!" While breeding from quality stock is certainly a good idea, single-minded focus on this practice leads to genetic bankruptcy. Imagine the genetic diversity of your flock as a pool of genes; each individual contributes to this pool. As you weed out undesirable traits, you reduce the size of the pool. When a superior individual, family, or strain is used too heavily, other unique genes are lost. When improvement breeding is taken to extremes, a breed or a species can become so narrow in genetic diversity that the animals become susceptible to diseases or reproduction problems can become prevalent — at this point, the population is so closely related that every one of them is a cousin or a sibling.

Diminishing Diversity

The best example of a small gene pool is the lack of diversity in commercial turkeys in North America. Today there are only two breeding companies that supply the genetics for most of the 280 million commercial turkeys produced annually — Nicholas Turkey Breeding Farms and Hybrid Turkeys. In 2004, there were three companies, each holding its own unique breeding stocks. That November, Nicholas purchased British United Turkeys of America (BUTA). Nicholas, which is itself owned by multinational poultry company Aviagen, compared its strain with that of BUTA and found the BUTA birds slightly smaller and less efficient growers. In a business decision, Nicholas sent all the BUTA turkeys to processing and then there were only two commercial strains in America.

Conservation Breeding

Breeding for the continuation of the population can best be termed "conservation breeding." This is the practice of making culling, selection, and mating decisions that maintain genetic diversity within the population. In practice, the simplest way to apply conservation breeding is to select the best individuals from each family or mating rather than the best overall individuals, which may have all come from one family or mating. Diversity should be valued within a strain or flock. It is this diversity that prevents inbreeding depression and allows the population to be mated for decades with no introduction of new blood. Conservation breeding need not be about using animals of poor quality or those that do not perform as needed; it simply requires that you retain animals from each family or mating so that none of these families becomes extinct.

A Philosophy of Good Breeding

Really good breeding begins with a philosophy that guides mating decisions. This philosophy is a way of looking at the members of the flock as only a part of the whole and weighing their value based upon what they contribute to continuation and improvement of the flock. In other words, the focus should be not on one exceptional individual, but instead to produce an exceptional flock that can sustain itself, and its high quality, for decades. Be a respecter of nations, not individuals. A good breeding philosophy is also about understanding the connection of the selection within the system of production, as well as selection for the desired product.

The Flock as the Focus

Once you see the flock as the focus of breeding, it becomes much easier to make decisions that will raise the quality of the flock while keeping the diversity high enough that new "blood" won't be needed for many decades. Such a decision is to limit the use of any particular male to one breeding season, with a few exceptional males being used for two or three seasons.

Another decision might be to keep hens for only two breeding seasons — this being a worthy consideration once the flock has reached the maximum population size for your enterprise. A last example is that of selling some of your best stock and keeping some lesser stock. This can occur when you have retained, for breeding, the maximum number of

RECLAIMING IDENTITY

The American Livestock Breeds Conservancy's (ALBC) *A Conservation Breeding Handbook* has a section devoted to developing a philosophy of breeding. The ALBC philosophy of breed conservation "is to maintain the breed's original identity, including its production, by using selection criteria that are consistent with the breed's history."

This sentence packs a lot into a small space. With this philosophy, the history of the breed, the techniques of selection used, the system of culture used, and the level of production are linked in order for successful conservation of the true breed type. What we need to understand is that a breed or variety is connected not only to the product it produces but also to the system of production, which will favor some individuals, and to the culling and selection techniques used.

siblings from that best family or mating and need to keep the best from the second or third family or mating to maintain diversity in your flock.

The Environment Is Essential

A good example of someone who applied philosophy to breeding very successfully is Dutch geneticist A. L. Hagedoorn. In his 1939 book *Animal Breeding*, Hagedoorn points out that the inherited genetic makeup of animals does not completely determine their productive potential—it only suggests their potential. He goes on to say that the environment in which the animals are raised can have an equal effect on the final quality of the animals.

Dr. Hagedoorn repeats that farmers put too much faith in qualities that can be seen, when differences may be largely environmental. It was his belief that selection based on productivity would naturally lead to the correct body type for the job and system for which the breed is designed.

Rather than focusing concern with the quality of any given individual, Dr. Hagedoorn looks at the line — assuming that many of the individuals within a line will reproduce similar offspring. He also writes that the best measure of the quality of an individual is observation of the type of offspring it would produce.

Putting Philosophy to Work

A good philosophy on breeding will help keep you on track with your goals. Start with a list of goals. Individual members of the flock should be valued for their contribution to the flock and to your goals, but with an awareness of limiting excessive impact to the genetic breadth of the flock. In other words, good breeding takes into consideration the sustainability of the methods of selection and improvement you choose.

Genes

Genes are the building blocks of a body, given by the parents. They are the "rungs" of the DNA ladder. A single gene is half the material needed to express a trait; it takes a pair of genes to express the actual trait and complete the DNA rung. Each parent provides half of this genetic makeup, and the combination of these genes forms the basis for the differences we see in individual turkeys. An ideal turkey is the expression of traits based on multiple combinations of genes.

Genes are either recessive or dominant in nature. A recessive gene requires that both parents contribute a copy of that gene in order for the trait to appear. A dominant gene does not require a matched pair from both parents — even when only one parent contributes a dominant gene, the trait the gene represents will be expressed.

Turkey genes are not well mapped at this time and it is not the purpose of this book to give a complete breakdown of turkey genetics, but to help the reader better understand the interaction of dominant and recessive genes and inheritance. In seeking to create an ideal flock of

HOW GENES WORK

To better understand genes, and to represent the contribution from each parent, we use letters to represent the genes and the complete gene pairs. Dominant genes are represented by capital letters and recessive genes by the same letter in lowercase. The letter combinations thus indicate:

Aa. A turkey whose pair of genes for a certain trait consist of a dominant gene from one parent and a recessive gene from the other parent. aA is also written as "Aa," as they both cause the same outward expression, and proper format is to give the dominant gene the first position.

AA. Inheriting a dominant trait from both parents.

aa. Inheriting a recessive gene from both parents. When a certain trait is described as recessive, only individuals with aa outwardly express the trait.

For example, in a flock of turkeys of one color, a different color could appear, even after several generations of breeding true. Such an occurrence could come from within a flock of turkeys, in which some individuals for the trait were AA and others Aa. Once two turkeys finally mate so that each are Aa, there is a 25 percent chance of an individual receiving the "a" gene from both parents — and out hatches a poult that appears different from its parents.

turkeys, what we are really doing is trying to bring together the multiple gene pairs that cause several characteristics to express; in this way, your turkeys could be plump, healthy, good foragers, and beautiful.

Breeds, Varieties, and Strains of Turkeys

We use different words to describe groups of animals based upon their different levels of similarities or relatedness. Within turkeys, we can say that the word *breed* applies little in that most turkeys share a common "type" (shape). This type defines what the turkey's body is designed to do — usually the production of meat. We could say that there are essentially two types of domestic turkey — the industrial and the heritage turkeys. While both are selected for meat, the industrial can no longer mate naturally due to leg and body proportions.

"Varieties" is a better word for the different races of turkeys. This word, in particular, applies to heritage turkeys, and can be used to differentiate between groups with varied color patterns or from different localities. We can say that all members of a variety of turkeys will share characteristics that make this population unique and identifiable. (See chapter 2.) Some examples of well-known turkey varieties would include: Bronze, Bourbon Red, Narragansett, and White Holland.

There are no registries for poultry, and there are no pedigree papers used to track ancestry. Instead, poultry are bred to a written standard that gives very specific information to ensure that the phenotypic traits of the variety are maintained. Phenotypic traits are what you see in an animal. They usually include color, such as Bourbon Red, but may include other specific information, such as leg or eye color. These traits can be easily seen and are good indicators that the unique genetic package of the variety is intact. In other words, purity of color pattern may indicate the presence of genes that convey certain performance traits.

In turkeys, the greatest value lies not in the perceived quality of a variety, but of a strain. Strains represent the work of one or more breeders over many years. A breeder's reputation is tied to the quality of his strain and the value of his stock is often a product of his reputation. Breeders often have much pride in the quality of their stock. This same pride, combined with a touch of peer pressure, tends to ensure that a master breeder remains ethical in his breeding decisions. The importance of the relationship of a master breeder with his stock cannot be overstated.

Producing a Strain

In trying to produce a quality strain of turkeys, you will note that certain individuals within your flock are superior in the qualities you desire. Why is this? Why is not the whole line of one even quality level? And why is it such a challenge to produce that one very special bird — let alone a flock of high quality? The answer is in its genes.

Think of your flock of poultry as a pool of genes. Within that pool are all the traits that can possibly be expressed by the offspring of this flock

LEARNING THE LANGUAGE

Finer distinctions of populations are important in our understanding of breeding turkeys. Let me clarify a few more terms before we go any further:

Strain. A group of somewhat related animals with a common heritage managed as separate flocks each under the supervision of a different breeder. A strain is the result of a bloodline being shared among several breeders. A group of birds under the supervision of one breeder is sometimes also termed a strain.

Line or Bloodline. A group of somewhat related animals with a common heritage managed as one closed breeding flock that is under the supervision of one breeder. If the breeder shares stock with other breeders, collectively the total population may be termed a strain, but only the original master breeder has ownership of the bloodline — the animals of the other breeders each are lines within the strain.

Family. A group of animals within a bloodline that share a common ancestor(s). A family is usually a subdivision of a bloodline, used to track relationships while managing the population. Most master breeders maintain three or more families of birds — this gives a source for quality new blood to use to prevent breeding individuals that are too closely related.

without the introduction of genes from another flock. In order for your flock to exist for any length of time, the pool cannot be so small that there is no variation; but neither can it be so large that only once in your lifetime will outstanding quality or specimens occur.

The nature of breeding is such that genetic diversity within a group of individuals must reduce over time, regardless of breeding methods used, unless every individual gets to reproduce an equal number of offspring. Balancing your desire for fast progress with long-term considerations such as the future health and viability of your line must be foremost in your mind as you set out.

In order to achieve success, your flock of breeding birds must have all the traits you wish to combine to form the ideal individual. Look over the flock. Does at least one individual have proper leg length? Are there birds with proper body size, feather width, eye color, fleshing, and other desired characteristics found within the flock? Any trait not expressed by your foundation birds will not spontaneously appear — that is, unless the trait is recessive and is carried by your foundation breeders.

Long-term success is measured not by producing one ideal specimen, but by increasing the frequency of expression of the hard-to-achieve gene combinations. Refining your bloodline increases the likelihood of the appearance of the traits you desire. It may also reduce the pool of genes too far. On the other hand, each time you add new blood you decrease this likelihood because you introduce new material that was not previously in your gene pool. Introducing new blood may also cause an overall reduction of quality due to less-frequent appearance of pairs of recessive genes. It is to your advantage to maintain the purity of your gene pool as you slowly refine it over years.

The old breeders always used to say that it takes 3 to 5 years to "settle the blood." What they were referring to is the refinement of the variability that comes from outcrossing unrelated strains. Quality in first-generation crosses of two strains, just as in hybrid vegetables, may not reproduce itself. By carefully selecting and mating the offspring of crosses we can, over a period of three or more generations, affect the genes of the population so that the population begins to produce predictable results. Most master breeders start by breeding one strain, and, after they learn how to properly breed, usually cross two or more strains to produce a population that best reflects their ideals. By refining your pool of genes, you can found a strain that is uniquely your own.

Breeding Methods

What if you do not have the patience to develop and refine your gene pool? What can you do to have excellent stock? The old answer to this is to simply find a master breeder for the breed/variety you like and periodically purchase new stock from them (usually new males). In this way you benefit from the skill of that breeder and by their commitment to hatching in large enough numbers to maintain quality. But success that is not based on your own ingenuity is not as sweet as the results from your own decisions.

What if you are just starting, and don't yet know the "secrets" of properly mating your birds? Again, the old advice is simple: buy from one successful breeder and keep the line pure. In this way, faults will slip more slowly into the bloodline while at the same time you learn to select and maintain the line. Plus, if you need new blood, you can go back to your source as long as they still are raising that breed/variety.

It is a fact that many of the great breeders of the past, and of today, usually crossed two or more lines to found and create their own bloodline. In almost every case, they worked with one line for 3 or more years before deciding what the line needed and where to go to get what was needed. The founding of the line usually did not coincide with the simple desire to make the line their own, but came after understanding the principles of properly selecting the breed or variety, and of how to mate the birds properly.

Today we have many great bloodlines that are dwindling. A sharp-eyed, young breeder could make a good name for him- or herself by forming a relationship with one of the older breeders of one of these lines, learning the proper selection and mating of the line from the master breeder, and raising enough offspring each season to allow the rule of 10 percent to apply (10 percent of a hatch will be of higher quality — thus progress can be made while breeding the line pure). As the young breeder grows in skill, areas of improvement for the line can be identified and rectified.

Cyclic Breeding or Spiral Mating

Once you have founded a line, the simplest way to breed to retain both quality and genetic diversity within the line is to use cyclic breeding. In this system, you set up three to five families of birds; let's call them families A, B, C, D. Females stay within the group where they are born. For example, family A daughters that are retained are added to family A. Sons retained as breeders rotate to the next group, so a family A tom is used to head family B hens (B to C, C to D, and D to A). This rotation would be one way to maintain a bloodline. George Shreffler of Ohio was the first poultryman to tell me he used this method. George said he learned it from Henry Miller and that it allowed him to go many decades without needing to add new blood. Cyclic breeding is also a common breeding management system with livestock (such as cattle and pigs) and is often referred to as **spiral breeding** or **clan breeding**.

In some cases, you may not rotate males for several years to fix traits within each family. You may also have pair matings within the families. Offspring of these would be added as above. So a family A tom mated to a family A hen as a line breeding, or even a family A tom back to his daughter or granddaughter, would produce family A females that could be added to the base family A flock and sons that could be used to head family B matings.

New Blood

Say you need new blood—whether for a particular trait that the bloodline is lacking, or simply because your start was of birds with some close relationships and you want to broaden your genetic base. If the same breeder who supplied the original trio is your source for new stock, then you are still within the same strain. If another breeder is your source and this breeder has birds from the same source you first used, and if the birds resemble the original bloodline, then you are still within the same strain. Adding new birds from this strain, you can more or less consider the same as creating a new family from within your own bloodline. In this case, you might call the new stock family B and your original birds family A and proceed by rotating the males of the now broadened bloodline.

If you obtained birds from another bloodline entirely, you could start a new family with these birds. The next couple of years would see an increase in the variability of the offspring of the flock, but after about 3 to 5 years, "the blood would settle" and overall quality would be more predictable.

Spiral mating is an excellent method to use when several breeders each can maintain only one flock of turkeys. Simply consider each breeder's flock as representing one family. By trading hatching eggs, flock A to flock B and so on, each breeder can receive the new males they need without the risk of bringing in disease by swapping birds.

Keep the Male Line Pure

Let's say in searching out new blood you decide to add birds from a completely different bloodline. In this case, you do not wish to start a new family, only bring in a small dose of new blood. There are a few ways to proceed. Most people blunder in, cross a new male over their hens and suffer in the next two generations because the cross has brought in recessive traits and broken combinations of genes that had expressed the good qualities of the original bloodline. The better way to proceed is to keep the male line pure and grade in the new blood.

Two Ways to Bring in New Blood

In keeping the male line pure, you avoid pitfalls by grading in the new blood in one of two ways.

Method 1: Cross a male of the new line to hens of your line. When you do so, save only the daughters for breeding and discard the first-generation sons. These "hybrid" females, in the following breeding season, are then crossed to a male of your original line. Although the offspring are now 75 percent blood of your line, many are still carrying genes or gene combinations of the new line, so variability may run high and impact your main line.

The preferred next step is again to keep only daughters and discard the sons: the daughters to be again mated to a male of your original line (which will produce birds that are 87.5 percent your line). At the 75 percent to 87.5 percent purity levels, the offspring of the new males and females can be used to begin a new family or can be used across your line. At the 87.5 percent level, you can essentially consider the birds to be pure members of your line.

Method 2: Use a male of your original line over females of the new line. From this mating, only daughters are retained. These are then mated to the same or another male of your original line. The offspring of the second cross can be retained (being 75 percent your line) if you like or you can once again keep only daughters (thus producing 87.5 percent). Starting with a female of the new line and grading to males of your line is the best method for introducing new blood.

Recumbent Reciprocal Selection

If crosses of the new line and your own produce outstanding specimens of either sex, you may want to maintain both lines pure so that

JUST LIKE MIXING DRINKS

A useful way to visualize keeping your male line pure or keeping your bloodline pure (they are the same thing) is to picture mixing drinks. Imagine that coffee stands for one bloodline and water another.

If I have a cup of coffee that is too strong, I can add a little water. If I combine one glass of half coffee and half water in equal parts to one of pure coffee, the result will be closer to coffee than combining two glasses that are each half coffee and half water. The first gives me 75 percent coffee and the second stays at 50 percent coffee and water. If I combine the 75 percent coffee in equal parts with a glass of 100 percent coffee, now I have coffee of 87.5 percent purity.

If what I want is a glass of water and I start with half coffee and half water, how much water needs to be added before the water will taste right? As long as I have pure water and pure coffee somewhere I can keep adding either until the result is close to right. If all I have is 50 percent coffee and water, I will never have pure coffee or pure water again.

you can periodically cross them to produce outstanding first-generation offspring. But don't fool yourself: first-generation sons of this cross are very, very unlikely to be good breeding birds. Maintaining two families to produce offspring to cross to each other is not a new idea; fanciers and the commercial industry have been doing this for decades.

"Recumbent reciprocal selection" sounds like a mouthful, doesn't it? Recumbent reciprocal selection is a process the commercial industry uses to produce faster growth and better egg production in poultry. The poultry produced from this process have the desired characteristics but will not produce offspring of equal quality. The key to this system is to find two or more families that when crossed produce superior offspring, and then to maintain both of these families pure as well as using them to produce the crossbred offspring.

Family Mating

In cyclic breeding, you maintain three or more families of turkeys. It is not necessary to rotate the males each year — a pause once in a while is not such a bad idea as it allows for a little distinction to form between families. Breeding from two members of the same family is family breeding. This method has some advantage, especially when many matings of a few birds are being used. One may breed an old tom back to the same hens for several generations — this would represent one mating within the family while the daughters are used under other toms. Mating within a family usually only occurs for one to three seasons before spiral rotations proceed as normal.

Line-Breeding

Developing and maintaining bloodlines is essential to obtain predictable results. The longer you breed quality poultry, the better your "eye" for quality becomes. Faults that you did not notice in your first few years are now challenges to eliminate. Eliminating inconsistencies in offspring produced now is the key to higher success. It is time to breed using a plan and that time-honored, indisputable principle — line-breeding.

There are many levels of relationships that can be utilized in line-breeding. Sire-to-offspring is probably the most classic example. This is

MASTER BREEDERS ON LINE-BREEDING

One winter I met an older cattleman, Gerald Fry, who was trying to educate people on the need and the principles of line-breeding. Much to my surprise, and later to his, I heard through his lips many of the same sayings the old master breeders of poultry have written or said to me:

- "Keep your male line pure."
- "Introduce new blood through the female."
- "Line-breed a high-caliber male to his offspring — for three generations — to produce sons that will produce consistent results when mated to unrelated females."
- "The male is more than half on any mating (if he is a line-bred male)."

It is interesting that traveling all over the world in search of consistent genetics (in grass-fed cattle) Mr. Fry found that real breeders are few and far between.

Just as in cattle, in poultry we know that real breeders are not the norm but are rare gems. Knowing how to mate a breed or variety is part of what makes a real breeder. But learning how to fix traits is what allows that breeder to produce top quality for decades. The single greatest tool for doing this is to reduce variability by reducing the gene pool through intelligent line-breeding.

often practiced for three generations of females, and the end result will be a very uniform group of offspring. To avoid inbreeding, males from such a mating are then mated to other females of the bloodline, but from a different "family." Breeder Jim Rines, Jr., of North Carolina advises that when a mating produces well (with high-average quality) there is no need to replace the hen(s) with her daughters; go no further with line-breeding for this family but save offspring to use in creating new families.

Another excellent cross is mating uncles to nieces across three generations. One of my biggest steps forward came in mating a male to three sisters, mating his son to two of the hens the sire did not "click" with, and then mating the grandson to the hen his father did not "click" with. This form of line-breeding is gentler and does not cause the same level of inbreeding depression that closer relationships cause. But it also allows you to embed traits shared by the three siblings used.

Another form of line-breeding that has worked for me is that of using a male back onto his granddaughters. This form is not so very exceptional at "fixing" traits, but is a good method of mating within a family while keeping inbreeding to a minimum. Best results would be obtained by using this method once a line has been founded and the variability has been reduced. Within a well-established family, I have used this method to produce outstanding birds.

Beware of Inbreeding

An intense form of line-breeding is the mating of brothers to sisters. This is inbreeding and in my experience is not a good method to use. Reduction of size has been my primary observation; some reduction in vigor has followed. Utilizing half-brother/sister matings is a method that Irvin Holmes, father of my mentor, Dick Holmes, found to be his best type of mating — but only when the two birds shared an extraordinary sire. I have not found this method advantageous. For the most part, I suggest you avoid brother-to-sister matings except under special circumstances; when you do choose to use it, be sure to outcross the resultant offspring in the following season.

Improving and Then "Fixing" the Males

A main point about line-breeding that I want to convey is that its primary goal is to produce males of superior breeding potential. The mothers of these males must be carefully selected over the generations, but it is the male offspring that will allow you to make the greatest strides in reducing variability.

Once you have a male of quality, take extra efforts to keep him alive for several generations to "fix" his traits onto his offspring. Three generations of breeding — to reduce the gene pool and allow the traits of the good males to dominate the line — will yield surprising increases in quality as long as you resist the temptation to add new blood. The more I study the subject of mating, the more clearly the importance of the male becomes.

Founding a Line or Strain

Another mentor of mine, Clint Grimes of Harmon, West Virginia, always felt a breeder should found his or her own line. The traditional method for doing this is to take a male of exceptional quality and mate this bird to females of various families, and lines, as one big flock mating. In some cases, some of the females may even be related to the male — such as half-sisters, mother, aunts, and so on. From among the offspring, select toms and hens and mate them together to form families of the new line. As the generations go by, you simply cross these families who all carry the blood of the old foundation tom. Diversity within the flock is acquired by the diversity of the original females used.

Rolling Matings

I first heard of this method in an article by Craig Russell, breeder of a number of poultry breeds and varieties and president of the Society for the Preservation of Poultry Antiquities. I later found the technique described as the "Old Farmer's Method" as well.

The method is simple. Each season, yearling toms are mated to old hens and old toms to yearling hens. After breeding, the hens are combined into one flock and the flock is culled down to the desired number of hens — same for the toms. The offspring of the two breeding groups can be toe-punched to allow identification. The best offspring of each mating is then kept and used as the yearling birds in the next season's matings.

The advantage to this method is you have only two matings and birds are mated according to age — young to old, old to young. The disadvantage is that it is hard to know exact relationships — thus there is a chance for some inbreeding to occur unmanaged. In fact, in such matings, about every conceivable line-breeding relationship is manifested. This form of mating does avoid brother-to-sister inbreeding, and most of the breeding will occur between distant relatives.

Trio and Pair Mating

"One tom and two hens a trio do make." Mating from three birds is not an unusual method for beginning in turkeys or for small flocks. It is best to start with three birds as unrelated as possible, yet with some shared points of quality. In some cases, this population of three is carried on over several generations. The tom may be culled the first year and replaced with his son. The hens would be replaced the following year by their daughters. After about 5 years, a new tom is purchased.

A trio can also form the beginnings of rolling matings or spiral matings. In rolling mating, simply breed the second-year young to old and old to young as previously described. To start a spiral scheme from a trio, in the second year breed the old tom to two daughters, one from each hen — breed one hen to her son from the tom, and breed the other hen to her son from the tom. After this, rotate sons and retain daughters as described previously.

A pair is simply one tom and one hen. Pair matings are usually only set up to: test the quality of breeding from a new strain; when trying to make fast progress in strain improvement through carefully selected matings; through line-breeding, producing males with prepotency of transferring good characteristics across females of diverse ancestry; or when managing a small flock as sustainably as possible.

HOLDING BACK A YEAR

One old method of mating turkeys was to hold the young toms for one season and mate from the previous season's young toms. In this method, the males are allowed to mature more before being used for breeding — a worthy consideration when hatching late in the season. In this case, all the hens go into one flock, the young toms are kept on a different part of the farm in a bachelor group, and last year's bachelor group joins the hens. While some of the hens will be sisters to the toms, the intensity of this form of inbreeding is reduced by the number of old and young hens being used. Toms are used for one season and hens for three or more seasons.

Stud Mating

Using one tom over one to twenty hens would represent a stud mating. The idea is that the sire of the offspring is known. Stud mating is often used as a component of another form of mating. It is not unusual to have several stud matings during breeding season.

Flock Mating

This is the method used by most hatcheries and industry. It comprises a single large flock of turkeys with several toms and many more hens. Dr. Roy Crawford suggests that a flock of 200 hens and 20 toms has enough diversity to avoid a genetic bottleneck. Hatcheries often breed their turkeys for one season, and then replace the entire flock at the end of the season. Industry often does the same. The advantage of managing in this way is that you avoid the feed costs of maintaining adult turkeys outside of breeding season. The disadvantages include potential loss of longevity in the flock — no turkey living for more than 2 years — and higher costs in raising the young stock as more must be hatched each year to provide replacements.

While mating from one flock is a viable method of breeding turkeys, if space is available, having more than one breeding group will better maintain diversity and give the breeder more control over improvement (through planned matings). But flock mating gives the best idea of the full range of character of the strain.

TOPCROSSING

The practice of topcrossing is that of going back to the same source for males each year. This practice allows for the production of high quality without the need for skill in breeder selection or culling. In some cases, this method is used to gradually move the flock as a part of one strain to that of another, slowly grading out the effects of the original source. One can also say that during the first few years of a spiral mating scheme, topcrossing is occurring to each family. Topcrossing produces birds with high levels of vigor and can result in an excellent flock of turkeys.

Wildcard Mating

Love that name! Sounds like a gamble that will surely pay off, doesn't it? During the early days of strain development, there may be many reasons to mate only a few select trios or pairs. Many birds may simply not have a mate that has the characteristics to breed well with them. In such a case, all the other birds must be put somewhere to keep them from interfering with the carefully selected matings. Many master breeders would save some eggs from this flock — and thus the wildcard mating came to be. In a well-bred strain, it is not unusual for the wildcard mating to produce as many quality individuals as the planned matings.

Outbreeding

Outbreeding is the practice of breeding unrelated turkeys. It comprises two forms of breeding: outcrossing (strains) and crossbreeding (varieties). In outcrossing, each season toms of an unrelated strain are sought and bred to the flock. In the first generation, hybrid vigor can be observed (fast growth and high immune function). But in the second and following seasons no increases in hybrid vigor will be observed. Quality will most often vary greatly in this form of breeding.

In crossbreeding, turkeys of different varieties are crossbred. Once again, hybrid vigor will be high in the first-generation offspring and quality will usually vary. Many old-time turkey raisers would crossbreed turkey varieties so that the color of their turkeys was different than that of their neighbors. In this way, a large flock of turkeys running completely at liberty could be sorted as to ownership when rounded up in the fall.

Faster growth, earlier maturity, and increases in meatiness may be observed by crossbreeding Standard/heritage turkey varieties. Good toms and hens must still be used. Some historic excellent crosses are: Bronze toms on Bourbon Red hens, and Bourbon Red toms on White Holland hens. Crossbreds should not be retained for breeding as variable results would ensue.

Outbreeding is usually done for the benefit of hybrid vigor and to give offspring that are easy to identify based upon coloration.

Breeding Tools

As we try to improve our strain of turkeys, there are some tools that help us achieve progress. Breeding tools guide decision making, measure progress, and help shape the methods of our success.

Principle No. 1: Offset Faults

This is the first and most important principle in breeding any form of livestock — simply mate animals together so as to offset each other's faults. When you mate two animals together that have the same faults you strengthen this fault. Or in other words, more of the offspring will have this fault.

Principle No. 2: Intensify Good Traits

The second principle of good breeding is to intensify good traits by mating together animals that share these same traits. This is really similar to the first principle but you are focusing on good traits instead of faults. By mating animals that share the same good traits, more of the offspring are likely to retain these traits.

Principle No. 3: Emphasize Vigor

Principle number three, select breeders for expression of vigor. The health of your stock and the future of your bloodline depend largely on the health of the animals you choose as breeders. Disease resistance, longevity, fertility, egg-laying ability, activity, and virility are all expressions of vigor. By making vigor a prime consideration in choosing breeders, you will be giving your bloodline the health necessary to survive well into the future.

Principles No. 4, 5 and 6: The Three Tests

The next three principles are test to prove the quality of a breeder. They are the pedigree, sib, and progeny tests. The Three Tests are qualifiers used to determine the value of any given bird as a breeder. Each of these gives some indication that a particular bird will likely produce well when mated correctly. In combination, they single out superior breeding birds.

The pedigree test simply is choosing your breeders with their ancestry in mind. What was the quality of the sire or the dam? What was the

INTERPRETING THE THREE TESTS

The aim in selecting breeders is simply to retain those birds that are likely to and then do produce the best offspring. It is easy to measure the quality of a given bird by observing it, but never confuse that with its breeding potential — only the Three Tests will reveal the true quality of a bird as a breeder.

Chicken Breeder Irvin Holmes of Maryland wrote that a bird qualifies as a breeder based on the ancestry and sib tests, but only remains a breeder based on progeny testing. A simple, but very important quality — remember this.

quality of the grandparents? A tom will likely produce daughters similar to his mother. A hen will likely produce sons similar to her sire. An exceptional bird from an average or poor line will not usually reproduce its like in quality. Breeders usually recommend using an average specimen from an outstanding line before an exceptional specimen from a mediocre line for this reason.

The sib test means that you consider what the brothers or sisters of a bird look like. A hen will likely produce sons similar to her brothers. A tom will likely produce daughters that are similar to his sisters. A bird with poor-quality siblings will likely not prove to be a good producer.

Progeny testing is the proof of the actual breeding quality. Looking at the offspring shows you what the bird in question actually produces. When an individual is bred to several different mates, but produces consistent results, you'll know its true breeding quality. No breeder bird stays in the breeding program unless it passes this test.

Principle No. 7: The Law of 10 Percent

Poultry are meant to produce offspring in high numbers — not unusual when you consider how many predators they have. When you mate birds in high numbers, you can observe a trend in overall quality: one bird in 10 will be a little better than the other nine. Ten birds in 100 will be better than the other 90, and one of these will be extra good. One hundred birds out of one thousand will be better than the other 900, but 10 will be

AVERAGE QUALITY VS. EXCEPTIONAL INDIVIDUALS

Average quality is the next concept you should consider. While it is important to use birds that produce those few superior offspring, it is equally important in making progress with your bloodline to use those birds whose average quality is very high. You see, some birds will produce a few offspring of exceptional quality and the rest will be of lower or middle quality. But birds that produce high-average quality of offspring will allow you to make progress with your breeding program. The ideal breeder is that rare bird that produces some superior offspring and has a high-average quality for the rest.

Now to contradict these principles a little, I suggest that most real advances in breeding come from those exceptional birds from poor bloodlines. If we think of any one bloodline and consider where it started and where it is today, we can ask ourselves how did those poor-quality birds transform into this wonderful bloodline? The answer is that the genes from the occasional exceptional offspring brought up the quality level over a period of years.

Putting this into practice requires patience to stick with a breeding program over many decades. So I say use those exceptional birds, but do not expect them to make sudden, lasting improvements for you.

extra good and one will be superior. This is the ratio of the distribution of high levels of quality. It is also a good law to keep in mind when you seek to make great strides in the quality of a strain.

Culling

The single most beneficial practice to better the quality and health of your flock is culling. To understand this, again consider that poultry are meant to reproduce in fairly large numbers. This may stem from the fact that so many things find poultry delicious. When we understand that these high numbers are associated with lower survival-to-breeding ratios, it becomes much easier to see that nature intended this prey animal to be selectively culled — either by man or by predation. And when you consider that with higher numbers come a few specimens of higher quality, the only question is how to deal with the bulk of the specimens produced.

An old saying is that the best tool you can use to improve the quality of your birds is an axe. This applies to immune function as well as type, meat qualities, or feather pattern. Dick Holmes, my mentor, used to tell me the story of a master breeder of White Leghorns who in his earlier years hired a poultry judge to come and cull his flock. The old judge locked himself in the poultry house and started catching and killing Leghorns. The story goes that the culls came fast and heavy. When he was done the breeder only had a trio left out of 150 birds. The breeder later commented that from that day on he made progress in his breeding. Not all of you will be willing to use an axe to dispose of unwanted and unnecessary birds, but you should still take away from this true story the lesson of the effects of good culling on the quality of a flock.

Many old books indicate that it is quite easy to breed resistance to some diseases. The basic formula is to use as breeders only birds that have gone unmedicated for the entire year. Cull all sick birds immediately (this happens to help prevent the spread of disease as well). Many poultry breeder friends of mine have noted that once they stopped masking the real cause of disease — a poor immune system — by discontinuing vaccination and medication and adopting culling, within 2 years the health of their flocks was better than ever. A year or two of putting away the medicine and bringing out the axe will do wonders for the health of your flock too.

Reducing the Gene Pool

Almost all breeding for improvement or consistency is about reducing the number of variables in the bloodline. The fewer the possible gene combinations, the more consistent the results. Inbreeding and linebreeding are two methods that produce indisputable results by reducing the pool of genes to favor select individuals, or rather, the gene combinations represented by the select individuals. You can use this idea to help make better use of those occasional superior birds from poor bloodlines

— but be sure that the reduction in your bloodline is not too severe as you must balance quality and diversity.

Skipping a Generation

Another principle to remember is that traits sometimes skip one or more generations. Sometimes a bird will produce offspring like one of its grandparents or great-grandparents. That is why it is so important to know the pedigree of your breeders. Knowing what they have behind them will help you to produce offspring of a certain quality every few years. It will also help you to understand that even though a superior bird did not produce offspring at the quality level you expected, the traits you want may still be in the "pool of genes" and may yet again surface. Do not add more genes to the pool by bringing in new stock; simply keep mating the line pure using the principles outlined in this section.

If He Was a She

Yet another principle that I find of the highest value is to look over a breeder bird and consider, "What would you look like if you were of the opposite sex?" You can find indicators of color breeding, feather length, width, and number, meat quality, even egg-laying potential by looking at the birds this way. Once you know what you are seeing, you will be able to cull your young birds for potential breeders.

Mate Only Those That Mate Well

To produce outstanding quality, I find it beneficial to be very selective on which males to use, to "see" the females of the flock as a pool from which to find mates for the males, and to put aside the need to mate every bird every year. I believe adopting this philosophy is the best way to start with average quality and improve the stock over just a few generations.

Start by selecting the male(s) to be used. Look them over carefully. Observe them. Notice and make a list of strengths and weaknesses. Now go to your flock of hens. Choose the hens that allow you to continue to maintain diversity within your strain, and choose females that do not share the male's fault but have some similar strengths. Mate only the number of birds that mate well together — even if a tom receives only one hen. If you

MATE 'EM ALL!

In some cases, you may not have room to keep birds that will not be mated; you may also have such demand for your stock that there is no such thing as an extra bird. In these cases, mate every bird you have. Improvement will come not so much from mating birds correctly, but from selecting quality birds. Each season, select young birds to join the breeder flock that represent the top 10 percent of each mating you made — do not keep more from one superior mating and fewer from the other matings or you'll lose diversity too quickly.

As your breeding flock reaches the maximum number of birds for your facilities, be more and more selective in both the young birds that join the flock and the old birds that remain for more than one breeding season. Good selection is more important than which birds mate to which birds, and it is the true basis for improvement of a flock of turkeys.

mate your birds with these simple ideas in mind, you will make progress quickly and you will be able to maintain diversity in your strain.

Be Hard on the Males

A young group of poults will usually comprise about an equal number of males to females. When selecting breeding toms from this young flock, keep in mind that you need only one tom in every ten. So from the first season of raising turkeys you can make progress by carefully evaluating the young toms and being very selective. Evaluate traits such as meat quality, feather pattern and color, shank length, vigor, dominance, appetite, and ranging ability.

As the hen flock grows in size, select females more intensively as well. But in the first few generations, be more concerned with meat capacity, egg-laying potential, body size, and health of the females. In later generations, include feather considerations.

Care of Breeding Turkeys

While many of you will purchase poults, keeping adult birds to be used as breeders is extremely important to have a secure source of stock of known quality. It is also beneficial in ensuring future generations of turkey raisers have access to their own stock.

Yards

Breeding turkeys need room for exercise and to satisfy their needs to explore, as well as to stave off any feelings of pressure that arise when no personal space is available. For a small mating of a single tom and 12 to 20 hens, a yard 14 feet (4.25 m) by 150 to 200 feet (45–61 m) will suffice. The yard can be of different shape if necessary when conditions limit the length, but it should be of similar area — roughly 2,100 to 2,500 square feet (195–232 sq m). With several such matings, it is advisable to have a similar space between pens. This can work well in that it allows half the yards to rest and turkeys can be rotated into these yards. In matings with more than one tom, the pens should be proportionally larger, with about 200 square feet (18.5 sq m) per hen. High grass and natural land contours will prove beneficial in preventing toms from interfering with mating, even if such contours prove more problematic in finding the eggs.

Use the breeding yards for only about 6 months of the year — from December 1 through June 1. The rest of the year, these yards can be used for other purposes on the farm, or grazed by other livestock. Avoid

TOE PUNCHING

Just as is done with chickens, toe punching can be used to identify poults from various matings. This gives the small producer the ability to keep track of close relationships, such as brothers and sisters, and it also helps identify which matings were most successful. A small producer who wishes to sell breeding stock to another producer can utilize the toe punches of the young turkeys to identify pairs or trios that are not closely related. (See page 86.)

DAY LENGTH

Birds can measure day length. Turkey hens measure the length of time from sunrise, or the time that the lights are turned on, until sunset, or the time that the lights are turned off. If you use natural lighting, calculate day length as being 30 minutes before sunrise to 30 minutes after sunset. (See Lights, page 213.)

Long day length is responsible for both **photostimulation** of reproduction and **photorefractoriness**, which includes and is characterized by cessation of reproduction. It is important to note that although these have opposite effects on the hen's reproductive system, they are both natural processes caused by long days. They each have important and beneficial effects on birds in nature (including wild turkeys), but photorefractoriness is a negative occurrence for domestic turkey breeders: it diminishes overall egg production by shortening the lay period.

keeping swine, sheep, or chickens here: the first two may harbor the erysipelas that may sometimes affect turkeys, and chickens may sometimes carry the cecal tonsil worm, which can lead to the disease known as blackhead.

In the summertime, maintain the breeding flock on range and feed a mixture of 40 pounds (18 kg) corn, 40 pounds (18 kg) wheat, and 20 pounds (9 kg) oats at the rate of 0.2 pound (0.09 kg) of grain per hen per day and 0.5 pound (0.22 kg) per tom per day. There should be enough space at the feeder that all the turkeys may eat at once.

Turkeys need ample shade that does not prevent ventilation. The roost location should be airy and shaded. It is best to remove the nest boxes after breeding season as this removes the possibility of the hens piling up in the same nests and suffering from heat prostration. Summer eggs are of little value anyway.

Winter Housing

Keep housing clean and dry. It should provide ample protection from storms (snow/ice/freezing rain), but may be open. Breeder turkeys should be allowed several hours of outdoor exercise even in the winter.

BREEDING BITS

One tom per 12 hens is a perfectly good number.

Because turkey hens like to lay their eggs near the front, nesting boxes should be about 15 inches (38 cm) wide by 24 inches (61 cm) from front to back.

According to a Michigan Station report of 1940:

- Feed consumption per hen turkey for 6 months per breeding flock is 34 pounds (15 kg) mash, 68 pounds (31 kg) grain
- Requiring 1.9 hours of man labor
- Yield 43.4 eggs per hen; 81.7% fertility
- Hatchability of fertile eggs 67.4%; hatchability of total eggs 55.1%
- Mortality of poults 10%

Outdoor nesting boxes can be used when breeder birds are rotated across pasture. This design includes trap doors that drop so that records of production may be kept for each hen – though such traps require that someone be on hand to let birds out throughout the day.

Lights

Turkey hens must be sensitized to light for a period when day length is relatively short, such as during the winter. Breeder hen poults are best reared so that the birds are 16 weeks of age or younger during the fall of the year. This provides the birds with decreasing day length or short days as the birds enter adolescence.

The optimal age for photostimulation is approximately 30 weeks for birds of heavy strains during their first year. If females reach this age during the winter or early spring, they will enter into the egg-laying period with good physical maturity. If not, they will lay small eggs for a shorter period. If breeders are kept for several years, they will cycle with the season unless artificial light is provided to increase day length as natural day length decreases.

Day Length and Productivity

Critical day length is the minimum number of hours of light that the hen needs to induce normal egg production. This is generally what is meant by long day length — that is, any day length that is longer than critical day length. Although 12 hours of light is generally considered to be a long day length and is stimulatory for turkey hens, 14 to 18 hours of light per day are needed for optimal stimulation for egg production.

Longer day lengths do not necessarily result in greater egg production. If day length is too long, the overall time in production can be reduced by photorefractoriness, or diminished response to long day lengths (see box on page 211). Day length should thus be just longer than critical day length. The goal is to maximize photostimulation and minimize photorefractoriness.

For most parts of the United States, 15.5 hours of light (artificial and natural light combined) is adequate to start hens in the spring. One 60-watt bulb for every 100 square feet (9.3 sq m) of floor space is adequate. You can use a time clock to bracket the natural daylight hours with artificial light. This often prolongs the natural lay period well into the fall. A typically adequate period of egg production could be anywhere from 20 to 30 weeks, depending on the birds and environmental conditions.

A proper sequence of short-day-length days ends photorefractoriness and restores photosensitivity in turkey hens. In nature, this is accomplished by the short days of winter, and utilizing these short days is probably best for small-flock producers as well. Hens can return to natural day length and rest during the winter. Hens need approximately 12 weeks for a proper rest period. The reproductive tract regresses, and the birds molt and become reconditioned for a new egg-laying period the next spring.

Approximately 8 weeks prior to mating, stimulate the toms with at least 14 hours of light per day. Do not let day length shorten for the toms or hens. Light stimulation is necessary for good sperm production from the toms and good egg production from the hens. Toms respond to light more slowly than do hens, thus their lighting period needs to be initiated before that of hens. Hens should be photostimulated with long days 3 weeks prior to the desired onset of egg production.

Storing Eggs

Eggs retained for hatching should be properly stored to give best results. A cellar or any other location with stable temperature and humidity will suffice. Optimal temperature for egg storage is between 50 and 60 °F (10–15.5°C); it's important to keep the temperature below 65°F (18.3°C) to ensure hatches with few stragglers. Eggs are best set within the first 10 days after being laid for best results, though good results will occur with eggs even at 14 days after laying.

Keep in mind — fertile eggs are either in a state of growth, though perhaps a very slow state, or they have died. So a fertile egg is "alive" at a microscopic level.

Egg Production

Egg production in turkeys can be improved in the same ways as it may be in chickens. Selection for capacity in egg organ areas, and good pelvic bone and pelvic/keel spread, are important indicators of potential production. Marsden and Martin offer some excellent advice in their book *Turkey Management* when they suggest there are four main factors worth consideration:

- Persistency of production (which can be measured by date of last egg laid)

ADVICE FROM POULTRYMEN

Penn State poultry husbandman P. H. Margoff found the following to be true: in early December, turning on the lights in the turkey pens at 4 a.m. resulted in production of fertile turkey eggs by February. He suggested using 60-watt bulbs. Like chickens, turkey hens begin to lay about 3 weeks after the lights are turned on. The intensity of light refers to the foot-candles of illumination produced by a lightbulb or the sun. Several breeder guides currently recommend 10- to 12-foot (3–3.6 m) candles as a minimum for breeder hens.

Nicholas Turkey Farm in Iowa gave hens a special laying mash, cod liver oil, wheat germ oil, and lights. About one month later, the turkeys would lay fertile eggs. The farmer kept 45 hen turkeys in pens about 15 x 22 feet (4.5 x 6.7 m).

- Length of pauses during season (due to broodiness or other factors)
- Rate of sexual maturity (age at first egg laid)
- Rate of spring production (amount of eggs in March and April)

Males have great influence on the production of their daughters and so should be selected on a basis of the production of their dams and sisters and for capacity of pelvic spread and pelvic/keel spread.

Broodiness

Turkey hens do go broody and will set on a clutch of eggs, but broodiness in turkeys is not as obvious as it is in chickens. When a turkey hen seems to spend a good bit of time on the nest, checking the state of her vent will help determine if she is in fact going broody. A broody hen will have a constricted and firm vent, with little moisture. Nesting without laying eggs, hissing often, and walking on tiptoes are all indications of the hen being in a state of broodiness.

When you wish to end broodiness of a hen and get her back into egg-laying, the basic principles are the same one would use for chicken hens — place the hen in a pen without a nesting area and preferably with a wire or slat floor. The idea is to prevent her from feeling hidden and to

> **PLANNED PARENTHOOD**
>
> To make sure any given tom is the sire of a hen's offspring, the male should be placed with the hens a minimum of 3, preferably 4, weeks prior to the collection of any eggs.

also remove the stimulation of warmth radiated back from the nesting material. Four days with cool air on her breast, without the ability to feel hidden, and with good mash and water will usually do the trick.

If under artificial lighting, hen turkeys will lay nearly all year (except during molt). One hen laid 251 eggs in 1940, but average production is closer to 123 eggs per hen per year, with the hens laying at about 50 percent rate of production during peak season (January through June).

Selection

If a turkey has good skeletal structure and sound legs with a balanced profile, the next most important section to pay attention to in choosing your breeders is the breast. By 1940, breeders and poultry judges placed

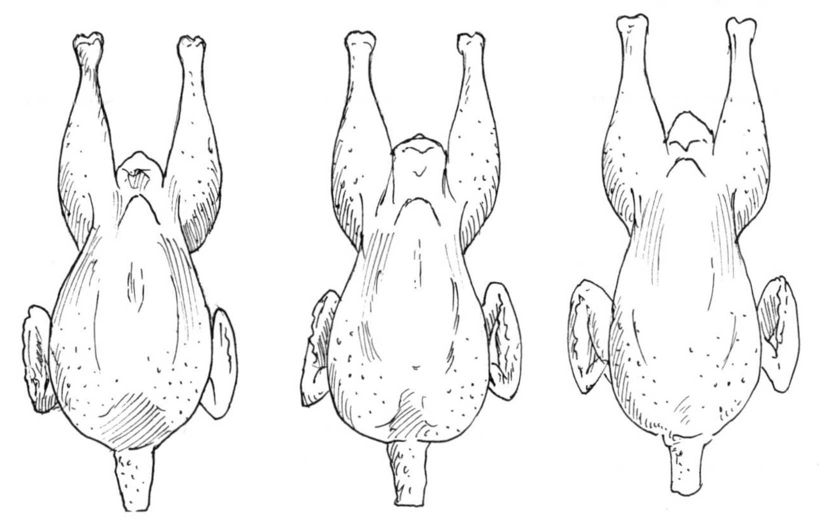

At first glance which turkey seems ideal to you? If you said the one on the left you need to look again. The left bird is wide at the front of the breast, but flesh does not carry back and keel does not extend between the legs. The bird on the right actually has the most flesh, with plenty of keel length. Such a bird makes for an ideal balance between fleshing and natural mating.

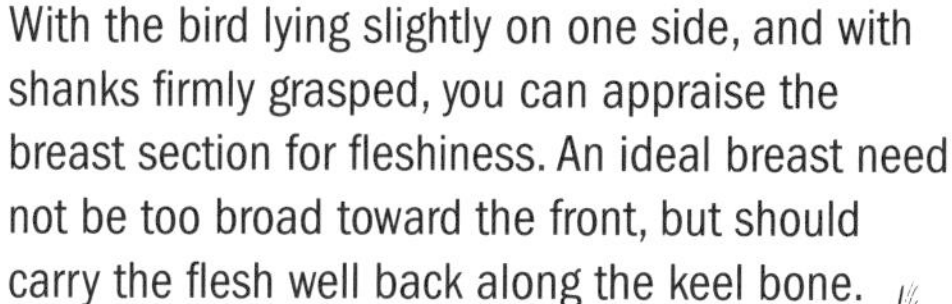

With the bird lying slightly on one side, and with shanks firmly grasped, you can appraise the breast section for fleshiness. An ideal breast need not be too broad toward the front, but should carry the flesh well back along the keel bone.

Heart girth is the distance between the wings and represents the internal capacity for heart and lungs.

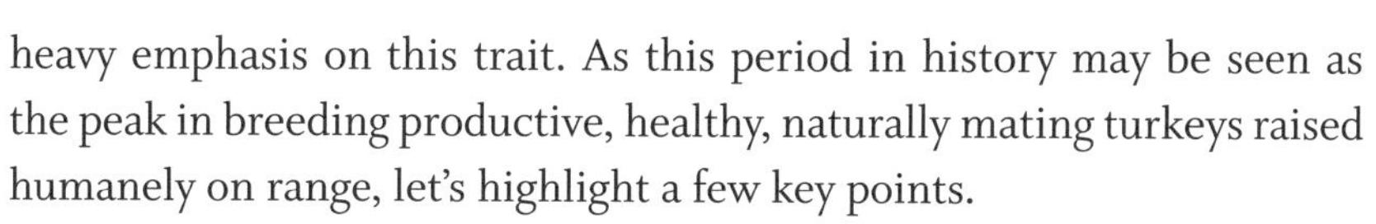

heavy emphasis on this trait. As this period in history may be seen as the peak in breeding productive, healthy, naturally mating turkeys raised humanely on range, let's highlight a few key points.

Width and length of the breast are important. While a long breast is desirable, potential breeder birds should have good width. Select away from birds overly wide in front, especially when such birds do not carry this width rearward. Select for extremely heavy fleshing across the whole of the breast and keel area.

The wishbone area should be very heavily fleshed, the flesh should extend over the spring of the ribs smoothly and flatly, and the rear of the keel should be blunt and not narrow — the flesh extending well back, leaving width between the legs of the turkey when the legs are held together as the turkey is examined on a table. The single most overlooked area of the turkey is the fleshing on the keel between the legs; if this area is well-fleshed, then your turkeys will have better shape and will produce more meat while still being able to mate naturally. Heavy fleshing at the rear of the keel bone, as well as the front and middle, should be your goal.

Avoid any turkey heavily fleshed in front and rear but lacking flesh in the middle of the breast and keel section. You also want a nice, deep keel bone to give the bird ample capacity for internal organs. The front tip of the keel should be well covered with flesh and have no tendency toward forming a "knob."

The back should be wide, as should the spring of rib, and it should be flat. When viewed from above, the body should appear rectangular and massive. Appraise the heart girth — the width between the wings. Width in this section allows more space for heart and lungs and will result in a carcass with more flesh.

The back, when a hand is laid on it and moved from front to rear of the bird, should be nearly flat. This is a sign of good skeletal structure. A bird that is curved of back usually has a taper downward from hips to tail (when felt). Such a bird will be found to have a less prominent breast and less internal capacity overall — it will have poor skeletal structure. Avoid legs that are bowed, crooked, or with enlarged hock joints. You want your turkeys to be able to mate and to range, neither of which can be accomplished with poor or defective legs.

A good method of maintaining the best balance of meat production and natural mating is to pay attention to the comparative sizes of the length of the keel bone and the length of the shanks. Shank length can be measured using calipers by bending the shank 90 degrees in relation

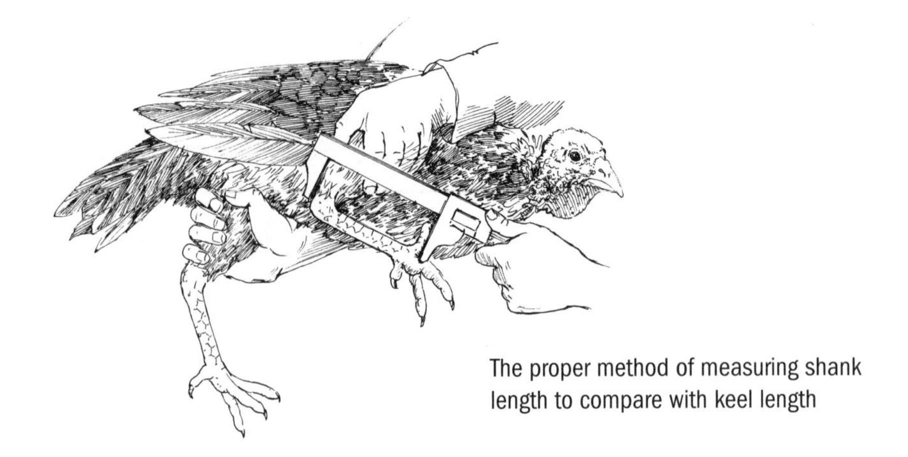

The proper method of measuring shank length to compare with keel length

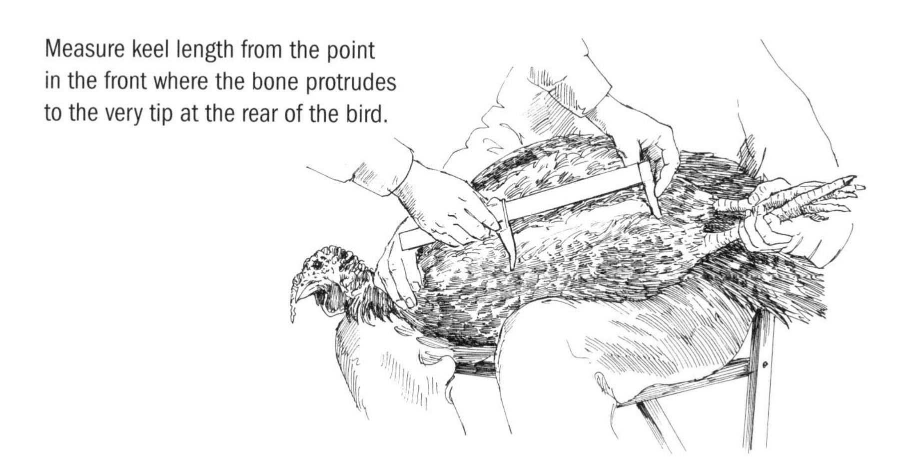
Measure keel length from the point in the front where the bone protrudes to the very tip at the rear of the bird.

to the drumstick and measuring from the back of the shank to sole of the outside toe. The keel is measured from the front of the keel bone to the rear tip. Both the keel and the shank measurements should be the same or very close, with keel bone being slightly longer than shanks by age twenty-eight weeks.

Turkey carriage is a good indicator of natural mating ability. Heritage turkeys tend to stand somewhat upright, with the breast carried a bit high. Turkeys that tilt forward cannot mate naturally. Turkeys that stand very upright can mate naturally, but have comparatively little flesh on their bones. Look for balance between fleshing and a structure that allows natural mating. I have found that a very good turkey has much the shape of a lowercase letter "y" turned upside down — though perhaps

TURKEY SADDLES

Turkey saddles are pads designed to fit over the backs of hen turkeys to protect them from cuts and abrasions during mating. They are necessary when a small number of females is being used with a tom, when hens have broken many back feathers and are still part of a breeding flock, or when you are trying to preserve a hen in best possible feather condition while using her for breeding.

Such injuries result from the male's sharp talons, the pointed tips of the spurs, and his weight. Trimming the sides of the toenails and blunting the spur tips will greatly reduce the need for saddles.

seeing a triangle in the body is perhaps the best way to understand outward appearance of excellence.

Wry necks do occur, often caused by *Mycoplasma iowae* (see page 265). Cull a turkey with a wry neck immediately. Consider submitting it to your state diagnostic laboratory to see if you have a disease to manage.

Applying What You've Learned

First, decide what your space limitations will be. How many turkeys can you keep each year? How many separate pens can you maintain? In any case, it is best if you can provide two or more pens — this will allow you to hold some turkeys or to set up two or more matings. Now choose the breeding method that will work best for your space. Flock mating will have some appeal due to the simplicity of breeder management. Keep in mind a goodly number of turkeys is preferred when using this method. Rolling matings will work well when all the turkeys will be used each year and when two pens are available. Holding a year will also work in a case of two pens. Spiral mating is the best method of managing a flock for many

SELECTION CONSIDERATIONS

- Look for fullness of breast and leg fleshing.
- Breast should be broad and full the entire length.
- Keel bone at the age of twenty-eight weeks should be as long or longer than the shanks (measure by bending the hock joint and then the toe, and measure from the bottom of outer toe pad to the upper part of the hock joint).
- Drumsticks should be plump with flesh extending well down toward the hock joint.
- Back should be wide at the hips with fleshing on the thighs.
- Heart girth should be broad; avoid narrowness here as it will constrict heart and lung capacity.
- Carriage should be upright with the breast carried above horizon; a breast carried low is poor and interferes in breeding.
- The gait of a turkey should be even and fairly smooth. Cull any turkeys with an uneven gait or whose tails wave left and right extensively as they walk.

generations. In some cases, you may not have three pens that can be used for turkeys. Here, two or more other local farmers may be willing to work together with you, each managing a flock that represents one family of the spiral — collectively, you and these other farmers will be managing a strain. If you are just starting, you may try sourcing or topcrossing or outbreeding. But once you commit to breeding these magnificent birds, you will need to think in terms of propagating them over generations.

My own experience has shown me that good culling and selection practices are 90 percent of breeding. The mating of the birds themselves is largely about managing the population. In other words, as long as you keep birds from each family, culling and selection have more to do with producing quality. So don't go chasing exotic breeding methods or stress over exactly which two birds mate well together; learn to select and cull properly and you will have a great flock of turkeys.

To have great success, choose a breeding method that fits your farm or poultry yard and will maintain some diversity in your flock. To improve the quality, hatch as many as you can, cull most, and select according to the selection considerations listed below. Your skill as

- The most important ages to weigh and compare young turkeys is at twenty-four and twenty-eight weeks.
- Young hens should lay a minimum of 45 eggs during any 13-consecutive-week period between January 1 and June 1. (That is a 50 percent rate of lay.)
- Ideal egg size is 3 ounces (85 g); select eggs from 2.7 to 3.3 ounces (76.5–93.5 g).
- Select for egg size during April and May of first year of lay.
- To select for rate of maturity, choose toms and hens at ages between twenty-four and thirty weeks that have the fewest pinfeathers and fullest covering of fat.
- Keep hens that have 80 percent hatchability of eggs (after candling).
- A ratio of 12 hens per tom is good for fertility. A lightweight tom can cover as many as 20 hens. Keep hens and toms with 90 percent fertility.

breeder will improve each year as you handle more turkeys, and you will develop an eye for the better specimens. I recommend you pay special consideration to the breast section — especially the amount of flesh between the legs, a blunt keel (due to flesh coverage), and coverage of the keel bone itself. If you do this and pay attention to the ratio of leg length and keel bone length, you will have a valuable flock of quality, naturally mating turkeys.

Artificial Insemination

This section was written by Leonard S. Mercia and contains excellent guidance for those who wish to use artificial insemination with turkeys.

If you're serious about producing hatching eggs, artificial insemination is the way to go for heavy turkeys. Heavy toms cannot mate naturally. Their breast is too large and their legs are too short, and they do not have the proper balance or agility. If your toms are of this type or body size, then you have to use artificial insemination.

All commercial turkeys are artificially inseminated. Hens are sometimes inseminated artificially to supplement fertility if natural mating has not produced good results. It is also a very good method to use to secure fertile eggs from very old or injured males. Artificial insemination also reduces wear and tear on breeder hens because they are not continuously mounted by the toms. The toms and hens are kept separate; semen is collected from the toms and then administered to the hens.

Collecting Semen

Semen may be collected from the toms two times per week. To work, or "milk," a tom requires two people.

1. One person places the bird on a padded table or a bench with the breast resting on the surface.
2. The primary person, or "milker," as he is known in the turkey industry, stands by the table or sits on the bench. The tom is in front with its head to the left and its tail to the right for a right-handed person. The helper holds the legs together and downward.

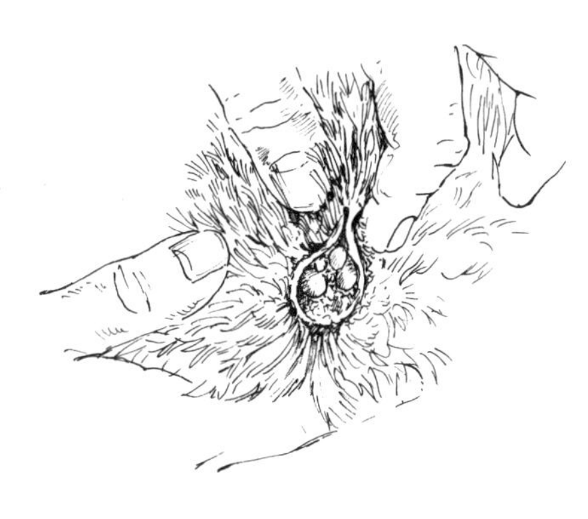

3. The milker then places the left hand on the back of the tail and the right hand on the ventral part of the tail or rear part of the abdomen.
4. The milker stimulates the tom by stroking the abdomen and pushing the tail upward and toward the bird's head with the right hand.
5. The male responds, and the copulatory organ enlarges and partially protrudes from the vent. If the copulatory organ does not protrude, the tom is probably not sexually responsive yet and needs more time.
6. If the bird does respond, the milker brings the left hand under the right hand and pinches off the cloaca at the walls of the vent with the left forefingers and thumb while the copulatory organ is exposed. The right hand is used to provide inward and upward pressure beneath the cloaca.
7. The semen is then squeezed out with a short, sliding, downward movement of the left hand and an upward pressure of the right hand. If you imagine a scooping action, especially with the left hand, it might aid the process.

Do not touch the copulatory organ during collection. Toms become trained quite quickly and ejaculate easily when stimulated. A tom produces 0.2 to 0.5 mL of semen per collection.

The semen should be milky in appearance and must be free of fecal matter. Some contamination can be avoided by withholding feed from the toms 8 to 12 hours before semen collection. Also, practice improves a handler's success rate in both handling the toms and collecting clean semen.

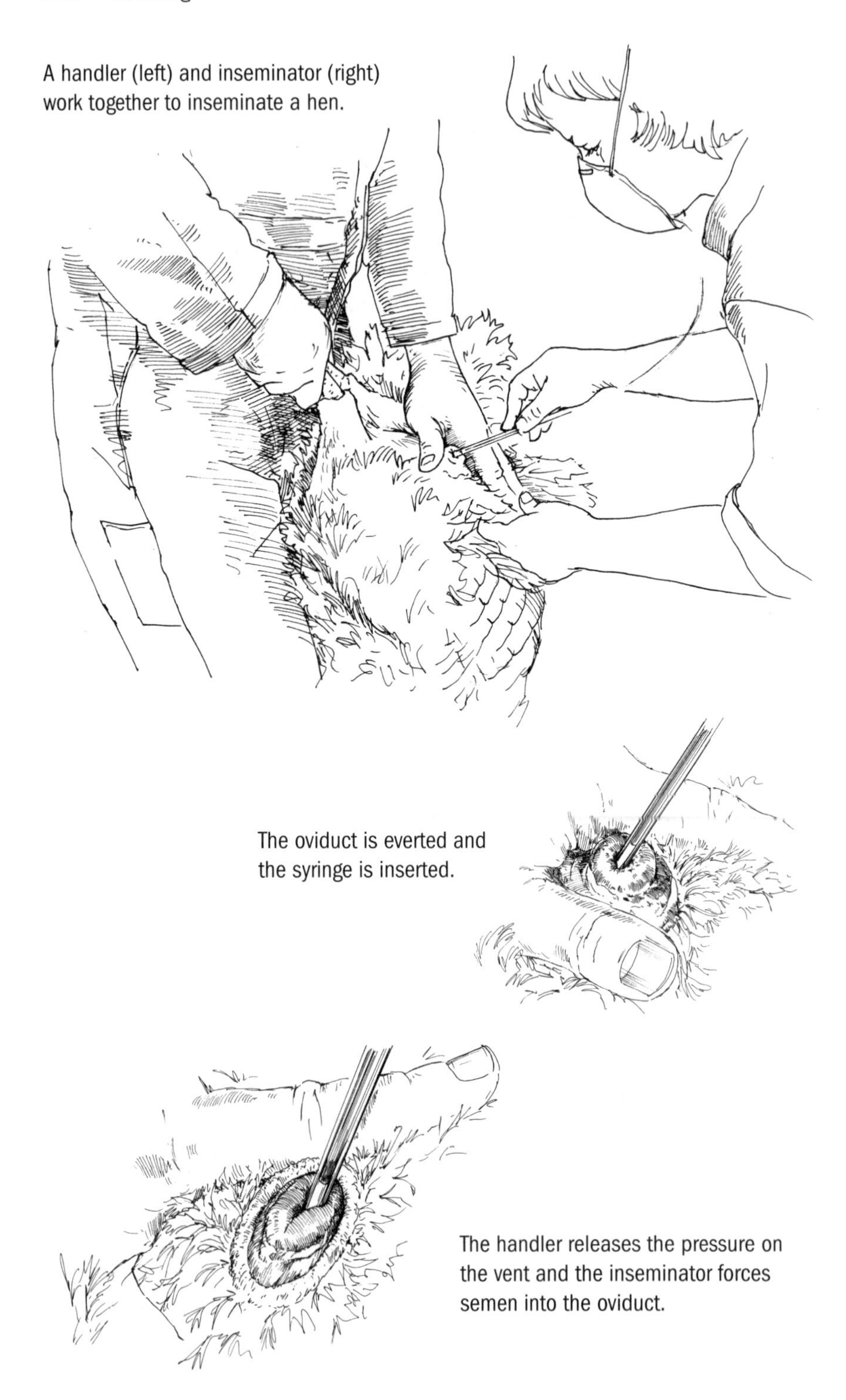

A handler (left) and inseminator (right) work together to inseminate a hen.

The oviduct is everted and the syringe is inserted.

The handler releases the pressure on the vent and the inseminator forces semen into the oviduct.

Collect the semen in a small syringe (without the needle), such as a 1 mL tuberculin syringe or a small, clean, dry test tube; glass beaker; or stoppered funnel. Unlike some types, turkey semen cannot be held long and ideally should be used within 30 minutes of collection.

Inseminating the Hen

Hens can be successfully inseminated with 0.05 mL of semen. However, inseminating an amount this small is difficult for small-flock producers to do without special equipment, which is expensive. Using an inexpensive 1 mL tuberculin syringe, you can administer 0.1 mL of semen, twice the recommended amount, quite accurately.

Good results are achieved when hens are inseminated twice within 10 days of egg production. Insemination should be done weekly thereafter for optimal fertility or least every 2 weeks for hobby flocks.

Again, insemination is best done with two people. The objective is for one person to evert the oviduct while the other person inseminates the hen. There are several ways to handle the hen to evert the oviduct, and each handler can experiment to find a suitable way. Two possibilities follow.

Everting the Oviduct: Option 1

1. Pick up the hen with both legs in the left hand if you are right-handed.
2. You may wish to lean back against a support. The hen's breast may or may not rest up against your knee, left or right.
3. Place your right hand on the tail of the hen so that your palm is to the right of the vent and your fingers are above it and your thumb below it.
4. With your right hand, press the tail of the hen toward its head to evert the oviduct. Rotate your hand away from your body. Using downward pressure with your thumb helps the process.
5. The oviduct will come into view on the left side of the cloaca; this is both the bird's and the handler's left.

Option 2

1. Pick up the hen with both legs in the left hand if you are right-handed.
2. Sit with the bird facing you, with the hen's breast resting on your lap.

BREEDING SOLO

Nineteen percent of female turkeys are able to self-fertilize their own eggs. This is called **parthenogenesis**. Few of these eggs hatch. The USDA's Experiment Station in Beltsville, Maryland, successfully hatched a few poults from an all-female flock some 50 or more years ago.

3. Expose the oviduct by exerting pressure on the abdomen while simultaneously forcing the tail upward toward the head. The oviduct can be exposed only in hens that are in laying condition.

Inserting the Semen

Once the oviduct is everted, the inseminator then inserts a small syringe without the needle 1½ inches (3.8 cm) into the oviduct. As the handler releases the pressure on the vent, the inseminator forces the semen into the oviduct and removes the syringe or tube.

Special devices, such as glass tubes or plastic straws, may also be used to inseminate. In this case, a rubber tube is attached to the straw. The straw is placed in the oviduct and the inseminator blows the semen into the oviduct as the pressure on the vent is released. Sperm can be stored in the oviduct for several weeks. However, fertility is at its peak during the first days following insemination.

◆ ◆ ◆

Breeding a high-quality flock of turkeys is not so difficult and is intensely important at this period in history — we simply have too few people breeding this rich genetic resource. Don't be intimidated by this undertaking, and don't feel that your enterprise will be too small to make a difference. Whether you can breed from a trio each year or a flock of 200, your efforts will help keep this genetic resource available for future generations. Uncle Tom Turkey needs you to survive.

You can do this. And for all your efforts, the turkeys will reward you with their time, interest, and antics. Who could ask for more?

11

The Turkey Enterprise

IF YOU WISH TO RUN A TURKEY ENTERPRISE, a little planning will ensure success and help avoid pitfalls. The best place to start is with an unbiased look at the various elements needed from start to finish. All of these components are necessary; any one can become a bottleneck that will prevent success.

The following are critical elements of a successful turkey enterprise:

- Consistency
- Source of stock
- System of production
- Feed
- Transporting
- Processing
- Product marketing
- Product storage and shipping
- Inspections and certifications

Let's look at each of these in turn.

Consistency

Consistency should be foremost in mind, as it pervades all aspects of a successful turkey enterprise. Without consistency, quality can vary such that the enterprise cannot be counted on to produce results with any predictability. Without a measure of predictability, consumers will be hard-pressed to understand the uniqueness of your product, bond to the benefits it offers, or promote it to others based on its merits — merits that will vary from unit to unit.

You can say that consistency identifies and defines the product. As you strive to succeed, you must understand the importance of consistency in each step of your enterprise.

Source of Stock

You have no enterprise at all, not even an inconsistent one, without a source of stock. If the business is to be seasonal (that is, you purchase and raise one batch per year and sell all), then you must understand that sources of that stock may vary greatly from one another. If your source of stock changes each year, you may have plump turkeys at twenty-eight weeks of age one year, and skinny birds at thirty weeks the next year. Needless to say, the second season could mean very disappointed customers and the end of your enterprise.

Likewise, your choice of breed/variety can have an impact on plumpness or flavor or consumer expectations. Breeds were created to produce predictable results from generation to generation, within specific production systems. There is a great advantage in having one source of stock of one variety/breed of turkey. The commercial turkey industry controls the quality of stock very tightly in order to add a great measure of predictability to the enterprise; you can learn something from the industry's example.

System of Production

The production system in which you raise turkeys will have an impact on rate of growth, tenderness, flavor, and nutritional value of the final product. It will also have a great deal of influence on your marketing efforts, as consumers will develop expectations based on the real and perceived value of the system or upon the results of the system.

The system of production will be dictated by a combination of available land and desired goals. You could conduct the turkey enterprise in conjunction with pasture management (in the case of a large parcel of

HOW TO CATCH A TURKEY

Capture the turkeys in a calm way. Turkeys are large and have powerful wings. If a turkey flaps its wings against a hard object — like a wall, crate, or another turkey — damage can occur, including broken bones and bruises, which may cause parts of the turkey to become unsalable. To prevent this trauma, walk turkeys calmly into a temporary penned area. In some cases, you may be able to walk turkeys into crates designed with doors on the sides.

In other cases, you will need to catch the turkeys by hand. The best method to catch a turkey is as follows:

1. Walk the turkeys into a smallish, enclosed area.
2. Gently but firmly reach down and place a hand over each wing of one turkey.
3. Press the turkey to the ground with its head facing behind you.
4. Shift the hand closest to the rear of the turkey to secure the legs.
5. Wrap the other hand around the turkey's breast with the turkey's head facing behind you.
6. Pick the bird up. It is important to maintain control of both the turkey's legs and wings as you carry it.
7. Gently release the turkey into your shipping crate, head first.

Remember, turkeys should always enter the crates head first and exit the crates head first. Practice this method and you will have few injuries.

land), with gardening efforts, or as a stand-alone enterprise (where fixed buildings and pens are used). Such factors as rate of growth and tenderness can be influenced by the amount of exercise received, which, of course, is controlled by the system of production. In order to produce the product your consumers know and desire, with any measure of predictability, you must be consistent in your system of production.

Feed

Perhaps no other factor can change the end quality of the product more than feed. "You are what you eat" is an old adage that can be readily applied to turkeys. High-protein feed results in faster rate of growth. Low-protein feed results in slower rate of growth. Feed content will influence flavor.

I knew a poultryman once who changed feed sources frequently, always chasing the best price; he never was able to produce high-caliber poultry with any consistency, though others using the same bloodlines produced superior birds. The simple reason for this lack of success was not genetic, but management — he did not provide consistent, quality feed.

I have also been a part of many organized tastings of various animal species in which breeds were compared for the flavor and texture of the meat. At these tastings, it became quite clear that while breed did make some difference, the diet of animals contributed greatly to the flavor of their meat. A diet with a good bit of corn seemed to coincide with the winning animals, regardless of species, in those tastings in which the animals were fed different diets.

Once you find a feed that gives great results, stick with it!

Transporting

The loading and transporting of live birds is an area where injuries and loss can occur. This can be traumatic for the birds and costly for the producer. Preparation with the proper techniques and equipment is the only way to ensure success.

Crate Specs

Crates for transporting turkeys should be high enough that the birds do not bump their heads, and low enough that they do not attempt flight; a typical size is 40 inches (101.6 cm) long, 23 inches (58.4 cm) wide, and 16 inches (40.6 cm) high. The crates should have solid bottoms so that when stacked, the turkeys in the higher crates do not defecate on the turkeys in the ones below them. When using crates without bottoms, you may add cardboard or use old feedbags to line the crates. Crates should never be overly full. Be sure the turkeys inside each crate have some room to move around.

Processing

Let's say you have purchased poults, brooded them, raised them on pasture, and fed them a good balanced turkey grower diet. In order to have a product to sell to your customers, you now need to process the birds. There are many things to consider here:

- Will you seek humane certification for the processing?
- Do you need U.S. Department of Agriculture validation to sell your product?
- Can you handle the workload of processing turkeys on your farm?
- If you need a processor, can he schedule you when you need the processing?
- Is there even a processor in your area or state open to the public?

Consider the processor your partner. In order for this relationship to work well, it is wise to understand that a processor does not stay in business the entire year based upon the work of processing your and other people's turkeys just prior to Thanksgiving. In other words, this sudden workload may be unexpected and he may not have staff in place or time available if he does not know well in advance of the processing date. Ideally, you should develop a relationship with your processor early in the season and reserve processing for a specific day. In this way, he can schedule the time and trained staff so that processing can proceed properly.

Processing Skills

Highly skilled processors are hard to come by and should be greatly valued. The skill of the processor, through all stages, will have a great impact on the end product you have to sell.

TRANSPORT TIPS

Turkeys need lots of air circulation during transport. Place crates on an open trailer such that the sides allow plenty of air circulation — do not transport large numbers of turkeys in an enclosed truck or trailer, as some will overheat or suffocate and die.

It is also a good idea to block the front side of the front crates during transport. The full force of the wind, as the turkeys are driven to processing, can actually make it difficult or impossible for those in front to breathe. This is why commercial trucks use trailers with a solid front, but are otherwise open.

If the processing facility smells clean, then there is a good chance of finding little bacteria on the surface of the end product. This will positively affect the length of time the product will stay fresh and the consumer's sense (through smell and even taste) of the value of the product.

If the processor handles the live and processed birds carefully, there will be few bruises on carcasses, few tears of the skin, and cut marks will be clean.

If the product is chilled correctly, then it will be tender and even juicy.

If the processed carcasses are bagged carefully, there will be little air to freezer-burn the meat and the carcass will have a pleasing shape.

Processing greatly determines the quality of the product that you offer for sale.

Introducing New Methods

Processors are businessmen and the running of their business requires that they work with a diverse group of producers of a variety of products. Turkeys are most often processed, following the natural order of the seasons, only in November or December to coincide with a perceived market centered on the holidays of those two months. Because turkeys represent only a small portion of the processor's operation, it is easy to understand that a processor may not have the time to research and learn new methods of cutting or packaging your turkeys since these will do little to alter his bottom line for the year.

A dressed turkey

If you desire that new methods or skills be used in the processing of your birds, you must be willing to learn these methods yourself, purchase any new equipment, and make arrangements to train the processor well in advance of the season. Even so, a processor may decide the new method does not benefit his business or justify any increased costs.

Processors are the turkey farmer's partners in producing a valuable product. The advantages and disadvantages of the chosen or available processor help frame the overall quality and identity of the product the farmer can offer. Producers must respect these important partners, without whom 6 months of effort and expense are for naught. Processors offer other advantages besides the basic service — they can be a good source of information on how to increase quality, as they see the results of many flocks. Form a good relationship with the processor, as good relationships are key for stellar service and assistance in producing a better product.

Product Marketing

Thanks to your processor, at this point you have converted your turkeys into a valuable product to sell. But do you have one product or many products to market? After all, turkey meat is one product, isn't it?

Turkey producers, and heritage turkey producers in particular, are lucky: in most cases, with few exceptions, they are operating an enterprise with only one end product. That end product is the whole turkey, ready for the consumer to roast. As soon as you divide the turkey into different parts, say breasts separated from thighs and legs, you have more than one product to market; you have turned one enterprise into two or more enterprises.

Let's say there's a restaurant that wants 200 turkey breasts — what will you do with the remaining wings, legs, and thighs? By cutting the turkey into many parts, the various end products each need a buyer. You will need to develop a whole marketing approach to sell each part of the turkey. I mention this, as it is a turkey enterprise's biggest cause of failure in its early years: taking one simple whole product and turning it into many smaller products, some of which are difficult to sell.

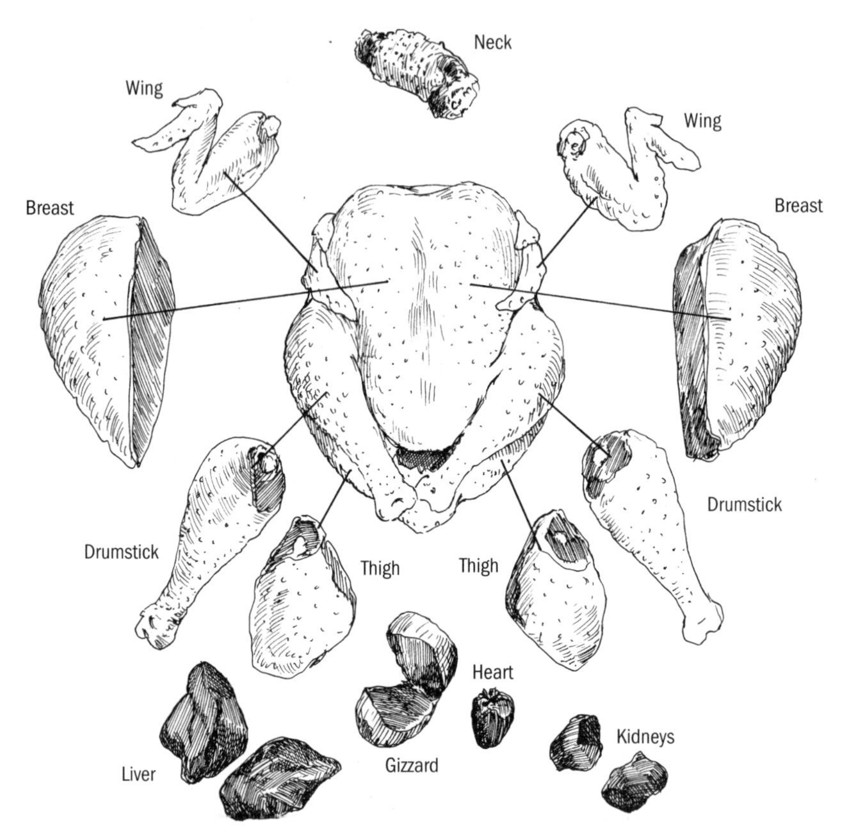

The whole turkey is the simplest product to sell. If you wish to cut it up, each part will need its own marketing plan.

Many farmers make the mistake of computing from their expenses the cost per pound and figuring this applies evenly across all products generated from the animals. Fact is, a farmer can get a better price per pound for breasts alone than for a whole turkey. But the thighs, legs, and wings are likely to sell for less than the cost per pound. So how can you sell part of the animal for less than cost?

Some parts sell for a premium, but the total dollars gained from those parts may not, of themselves, equal or exceed the total cost per bird. In other words, if you figure the cost per pound for raising that bird and multiply that by the weight of the bird, you have the total cost of raising that bird.

If you sell the breast of that bird at a premium, and these dollars do not exceed the cost of the bird, then you have not yet made a profit on that bird. The remaining dollars you receive for the thighs, legs, and wings should cover the remaining expenses and leave a profit — if you are able to sell these parts. You can brag to your friends of the colossal price per pound you got for the breasts, but what you might be leaving out is that you lost money as parts remained unsold and you have yet to cover the total cost of raising the turkey.

At the end of the season, there is also a possibility of having some whole turkeys left unsold. Remaining unsold products, even whole turkeys, represent an investment that has yet to return any dollars. You should be glad to sell these out-of-season at a discounted price, as every dollar they bring, assuming all expenses have been met, represents pure profit. In the words of my grandmother, a former turkey producer herself, "When you get paid, you first pay off all your bills; then, if you have anything left over, what is left is yours." True profitability of the enterprise is the difference between total dollars in sales versus total expenses.

Your goal for the first few years should be focused on producing one product and producing it well. Once you have made a success of that effort, you can begin to consider "adding" products to your offerings. Remember, some people like white meat and others like dark meat, and still others will buy the whole turkey. Start with the whole turkey as your product — consider carefully before selling the individual parts separately.

Product Storage and Shipping

If you have the ability to raise, process, and market your turkeys then you are blessed, but the timing of each element of the turkey enterprise is critical to success. How many producers have thought beyond source of stock, farm infrastructure, and processing to the timing of delivery and possible storage needs for the product? Ideally, you would time your turkey processing for delivery as soon after processing as possible. But you may need to store and transport the product, and maintain a safe temperature while doing so.

In some cases, the processor will be willing to work with you to provide storage and, sometimes, shipping and handling of the product as well.

If so, all fees, quantities, and timing limitations should be well-known to both parties long before the start of the enterprise. In other cases, you may need to purchase or rent equipment, such as coolers and refrigeration units.

Be aware that refrigeration devices can fail. A friend once transported his entire harvest of processed hogs to New York City in a rented refrigerated truck and did not notice the unit had a failure until he arrived at his destination. The result was a complete loss of product. Turkey meat is a perishable product. Realize the delicate nature of perishable products and have a plan to manage the products safely and correctly. Having a backup plan or experience in operating new equipment is also a good idea.

Direct Shipment

Many producers these days opt to ship their products directly to consumers. If you decide this is something you are interested in, you will need a shipping contractor such as USPS, UPS, or FedEx; boxes; insulation for the boxes; a cooling agent; and the ability to print labels and manage the database for your sales. Depending on the volume of boxes you will be shipping and your location, any of these shippers could turn out to be the best choice for your enterprise. Long before shipment, decide which offers the best price and benefits (sometimes boxes will be included in the price) and set up an account so that you can easily print the labels according to the shipper's specifications. If boxes are not a part of the shipping cost, then you will need to purchase your own boxes — once you are large enough, you can contact box companies to have your own printed.

SAFETY FIRST

In many states, you will need to take a class and become certified with a meat handler's license in order to legally transport your product. The class associated with this license is offered infrequently, possibly in the spring or summer. Check with your local department of agriculture or county Extension agent to attend. You will learn much about safe meat handling, so do not think of this class as a waste of time.

You will also need insulated liners for the boxes and cooling agents. Purchase these after a little research on the Internet. My favorite liner is an inflatable one because of the ease of storage — you inflate it just prior to using it. But there are also liners made of "green" materials and others of foam. Cooling agents come in a variety of forms as well. Some require immersing in water prior to use. All require that they be frozen for some period of time prior to use. Inflatable space fillers are nice additions as well, as they help keep the product from rolling around the inside of the shipping container and becoming bruised or damaged.

You will need access to a computer and printer if you will be shipping a large number of orders. About 30 to 40 orders could be handwritten, but anything more would be a waste of time better spent and possibly needed to fulfill and package orders.

Until the product is sold, it must be stored in a refrigerated state somewhere. From that location, it must be transported in a way that conforms with state and federal laws and protects the product from spoilage. Your customers should receive a product that makes them feel good about their purchase and offers them a unique experience.

Inspections and Certifications

The next consideration is proper inspections and certifications. Once you have developed your enterprise to any reasonable size, you will need proper inspection in order to sell your products.

USDA

I recommend seeking USDA-inspected processing if it is available near you. With this, you will be able to sell your products across state lines and to restaurants. It will be a validation from the U.S. Department of Agriculture that you sell safe products, and could contribute to your long-term success. If you start your enterprise by following all procedures to be USDA certified from processing to delivery, then you will have few roadblocks as you grow the enterprise.

Some folks will find it more satisfactory to do on-farm processing and have consumers pick up their own birds. In most states, you can legally do this, sometimes even selling direct to restaurants that will cook and resell to the public. But there are limitations to the number of birds per year that can be sold this way, and you cannot send your product across state lines. Contact your county Extension agent or state department of agriculture for the governing rules for your state. Many farmers start with on-farm processing. Just be sure that if you choose this route, you have clean processing facilities and appropriate storage for the processed birds.

Mobile Units

In North Carolina and elsewhere, such as Martha's Vineyard and others in Massachusetts, Kentucky, Pennslyvania, New York, Vermont, California, Washington, and Montana, farmers are now using mobile processing units for on-farm processing. Such a unit is made from or set on a trailer that can easily be towed to a clean site on the farm. It requires a source of clean water and electricity to operate but has tables, a scalder, and a plucker. Seek state or USDA certification if you design such a device.

Labeling for Market

Some farmers seek to meet humane, organic, or heritage guidelines, among others. Each type of certification has a set of requirements that the farm or farmer must meet in order to quality and in order to use such labeling. Often an inspection is a part of the process, and sometimes it may take several years of conforming with the guidelines to become validated.

These certifications represent marketing tools you can use to communicate the special merits of your farm and its system of production. If you decide to pursue any of these, first read the rules and understand what, if any, changes you will need to make to your current farming methods. It is better to start slowly rather than jump in and then out of the validation process. So choose with care.

Heritage

"Heritage Turkeys" is a term that has been defined by the American Livestock Breeds Conservancy to represent turkey varieties that are naturally mating. The importance of this term to the genetic diversity of domestic turkeys is enormous.

MARKETING BOON

In 1997, the American Livestock Breeds Conservancy conducted a census and found only about 1,300 breeding heritage turkeys remaining. At about the same time, Master Turkey Breeder Frank Reese, Jr., of Kansas was seeking a way to prevent the extinction of these turkeys that his mentors, and generations of Americans, had worked so hard to improve and utilize. Enter into this equation a group of people interested in promoting and sampling rare gastronomic and cultural treasures — Slow Food USA. Turkeys were grown, processed, cooked, and sampled — eureka! Intensive flavor was rediscovered. Heritage turkeys now have a market which has given the population a much needed boost.

The term "Heritage Turkeys" (see page 295 for official definition) is now approved for use on labels by the U.S. Department of Agriculture and requires no certification process. This means you can now identify your product for its uniqueness and merits. Economically, this means you can ask for a fair price for a turkey that takes more time to reach maturity and produces a superior product. This is a great boon — not only has the USDA recognized the difference, but so have some top food writers:

"We were amazed at the white meat, which was finely grained, extremely succulent, and the most flavorful." — Leslie Brenner, *Los Angeles Times*

"It has the turkey flavor that is merely hinted at in supermarket turkeys." — Marion Burros, *The New York Times*

"Now I understand what turkey was like before the triumph of the Broad Breasted White, and why turkey has once been considered a great treat — heretofore one of the mysteries of life."
— Michael Pollan, author and journalist

Currently in the United States, there are only two companies that maintain breeding flocks of industrial Broad Breasted White turkeys. That means the entire remaining diversity of domestic turkeys can be found in backyard flocks all across America, and is represented by the varieties, variety crosses, and color offshoots recognized in the American

Poultry Association's breed guideline book *The Standard of Perfection*. These "heritage" turkeys not only mate naturally, but they grow more slowly and thus produce meat that is more flavorful and requires slightly different cooking techniques to remain moist. (We could actually say that commercial turkeys, as a result of growing too quickly, obtain too much of their weight by water retention.)

"Heritage" as a term identifies a unique culinary experience connected to specific traits of livestock and poultry that are a part of our heritage. These products provide us with flavorful experiences that reconnect us with our ancestors.

Humane

Several organizations certify for humane farm practices, each with a specific focus, support, and requirements. Most of us will raise our animals and poultry in humane ways, and being validated for this good work makes much sense. Before you leap right in, however, be sure to see if the methods described fit your farm's methods and what, if any, expenses will be incurred to purchase new equipment to achieve compliance.

As an example, one of the big organizations requires the use of an electric stunning knife for killing poultry. While this device is excellent and has good science behind it, it is also a bit expensive for a small producer to purchase. If you plan to use a processor to handle that aspect of your enterprise, compare the processing requirements with the capabilities of your processor. In some cases the processor will not be able to offer the equipment required by the certifying agency.

Organic

Nearly everyone has heard the term "organic," recognized for use on labeling by the U.S. Department of Agriculture. It means using land, feeds, and animals in ways that reduce or eliminate the use of petroleum-based chemicals. The term originated to describe farming methods that promote the soil and its organisms and, as a result, produce products with more real and perceived health benefits. Farmers who wish to achieve organic certification may do so following some simple procedures (see

GETTING CERTIFIED

A turkey farmer wishing to produce organic turkeys for market must first seek an organic certification agency and make his application. Requirements include:

- The land must be free of prohibited substances for 3 consecutive years. The land could have been sitting fallow, and it need not have been under current ownership for that period. Soil testing may be used or required for validation.
- Turkeys must be managed organically from their second day of life onward. Poults purchased from nonorganic sources can qualify.
- Turkeys must be fed certified organic feeds.
- Only organic-approved medicines may be used.
- Animal structures must meet requirements for square footage, outside access, and portals.
- Stocking rates, even on pasture, may not be exceeded.
- Processing must be by an organic certified processor, with procedures in place to prevent the comingling of organic and nonorganic products and equipment.
- Documentation of inputs and procedures must be maintained and available for review.
- Fees must be paid.
- There will be an annual inspection and the possibility of "surprise" inspections.

box). Be advised: organic certification comes with fees, both for the application and annual inspections.

Certified Organic on a Smaller Scale

Producers who are small in scale may sell turkeys without certification. That is correct: a farmer following organic practices may sell up to $5,000 worth of products and label them "organic" in nature. While you would not be able to use the USDA's organic label, or that of any of the organic certification agencies, you will be able to print "Organic" on your products and mention it in your ads. This allows you to share the benefits of your farming methods while you grow the enterprise into a scale where certification becomes affordable.

If you decide to advertise organic products, be sure you are in compliance with the requirements and save all your documentation. This will make validation of your methods easier and may help later in the certification application process.

Marketing Your Turkeys

Marketing is absolutely instrumental to the success of any turkey enterprise. Let's say you have grown, processed, and stored a superior product, and you have your shipping methods and materials all lined up — the question now is, who is going to purchase your product?

Marketing is really a term for a specific form of human communication. What we strive to do is communicate what we have to offer, why it is useful or desirable, what makes it unique, and how it may be purchased. You don't need a degree in marketing to do a good job, but you do need to clearly communicate with your audience. So let's take the mystery out of this form of communication and look at some specifics needed to do a good job.

Tell Your Story

The first thing to understand is that you are not in the business of selling turkey, but of telling your story. The grocery store carries turkeys that are larger and less expensive. Why should a consumer buy your turkeys? It is because of your own unique story. Take a look at your enterprise and understand what makes it unique. Some of the things you do may even seem commonplace, but to others it may be exactly the sort of thing they wish to support.

Your story could be that of being a family farm. It could be one of a son or daughter returning to the farm, or of striving to save the family farm. Your story could be that you are striving to save rare breeds. It could be that you raise your turkeys on pasture, even organically. Your story could be one of striving to choose the most flavorful foods, or the most healthful. It could be a story of overcoming challenges or limitations. Your story could be one of following your faith. It could be a story of trying to help change the food production system.

SHARING YOUR VISION

Some questions to get you started:

- What is unique about your geographic location?
- How long have you owned your farm?
- Has it been in your family for generations?
- Is your farm a historic part of your community?
- Why did you select the animals you chose?
- Why did you select the feed you chose?
- Why did you select the farming practice you plan to follow?
- Who are you and your family? And what led you to start this enterprise?
- What is unique about you or your family?
- What is unique about the product(s) you will offer?
- How are you different from your competition?

Once you have these notes you will begin to see what makes you and your farming enterprise unique.

Here's a step-by-step method of clarifying your story.

Put it in writing. Sit down with a pencil and paper and jot down some notes about you, your farm, your farming methods, the animals you intend to use, and the goals and vision for your farm enterprise (see the Sharing Your Vision box above).

Organize your thoughts. Now sort this information into two lists. List 1 contains the obvious terms that consumers will be aware of, such as organic, heritage breed, pasture-raised, soy-free, humane, locally produced, USDA-inspected. As you form your marketing approach, these items will quickly convey a part of your story and communicate large ideas consumers may be looking to support. These terms do not go far into the heart of your own unique story, however.

List 2 contains those items that are unique to your particular farm, to your family, to you, or to your challenges. These are the ideas that are hard to express in one word. They will have messages like:

- Fifth-generation family farmer
- Son/daughter choosing to return to farm/farming after college

- Exceptional food flavors and/or textures
- Attempting to farm in a more sustainable way
- Saving rare breeds from extinction
- Overcoming a disability
- Following our faith
- Serving more healthful food to other families

These items will be the ones most unique to your enterprise.

Prioritize. Now take your two lists and place the items in each in order of importance. Which items are topmost for each list? What is next most important to you? Continue until you have each item in the order you feel it should appear on your list.

What's unique? Next, place an asterisk beside each item you feel is particularly unique compared to your likely competition; if some items warrant it, place two asterisks. When you are finished you may find that an item placed low on the list, meaning it was of less importance to you, is actually something very unique.

Second opinion. Ask a friend to review your list and highlight the top two or three items of importance, in their opinion. Allow them to add an item if they think of one that is missing.

This exercise should help you to begin to understand what your unique story is, and to see it from other people's point of view as well.

Marketing Principles

Before you begin writing your unique story and designing your marketing materials, look at a few principles you should have in mind as you do this. I find these seven principles most useful for all marketing tools and writing.

1. **Emotional content.** The marketing material should elicit a positive emotional response when read. Nothing stirs people to embrace an idea or set out to act on their feelings like emotion. Though negative marketing can induce fast action, positive marketing always works best in the long run as it brings with it the desire for continued results and builds the marketer's reputation.
2. **Personal connection.** The marketing must show the human side of the marketer and make a connection to people. People want to support people, not corporations. We instinctively like to form relationships,

and once we feel a personal connection, we have bought in to the idea or product and will wish to support that idea or product.

3. **Epiphany.** The marketing should have some information that is educational or new for the reader. New information intrigues our sense of curiosity and exploration. It also appeals to those who like to be the first to adopt a new product or idea.
4. **Uniqueness.** The marketing must demonstrate clearly the uniqueness of the product or enterprise. We want people to demonstrate that our product has value equal to or exceeding its cost due to its uniqueness. We want the consumer to understand that a product like ours can only be found here.
5. **Call to action.** The marketing should tell the exact action to follow to acquire the product. A call to action can be as simple as "Call Today" or "Order Now." The importance of this little bit of info is to plant the seed of acting now while telling them exactly how the product can be ordered.
6. **Focus.** You cannot be all things to all people, so don't try. Simply communicate what you feel is most important. If you try to promote every little thing you do, you will only confuse your potential consumer and they may leave without being able to remember your prime message. Just be yourself and communicate what is important to you. Keep your message short and simple, and people will understand what you are trying to convey.

MARKETING RULES FOR SUCCESS

- **KISS.** Keep it simple, stupid.
- **Bang for the buck.** Look at the cost compared to the number of times your promotion will be viewed or how long it will last.
- **Appeal.** Use marketing items that evoke a happy or positive response.
- **Poultry people are cheap!** Whenever possible, use free forms of advertising to target large new audiences — the best of those include interviews in local or national media.
- **Word of mouth.** Your best spokesmen and promoters are satisfied customers. Never forget this, as happy customers grow your business and your reputation.

7. **Consistency.** All marketing materials should carry the same message, though some will do so in more detail. Good marketing is the result of repeating the same message hundreds of times. Just the act of repeating the same message to a consumer lends credibility to the message — shows you are not simply "saying anything" to make the sale.

By applying these principles to your marketing efforts, you should be able to clarify what makes your story unique and write a short blurb about your farm or products. You will also have a short list of important bullet points that you can refer to as you create new materials. This list will be important to ensure your message stays consistent, regardless of whether it appears in short or long formats. Example: essential elements will be the same on a business card or a brochure, though on the second you will elaborate the message further.

Marketing Tools

There are many ways to get your story across once it is identified and written. You can use baseball caps, T-shirts, jackets, scarves, buttons, brochures, business cards, letterhead, signs, labels, mugs, posters, bumper stickers, decals for kids, billboards, local radio, newspaper ads, newspaper stories, vehicle lettering, key chains, or even flyers. Each of these tools will appeal to certain people, and thus be more effective in communicating your message to them. To use your tools effectively, think about what you like and what your likely customer base will find appealing.

Contact Is Everything

When you first acquire a customer, be sure to get their complete contact information — name, address, phone number, email address — to add to a database of customers. This list of people will represent your best goodwill ambassadors and promoters; best of all, they work for free. Not every customer will bond with your products, but be sure to address any issues and call people by name if you need to contact them should an issue arise with their order. I like to say, "Treat every customer as you would want your grandmother treated."

Getting the Word Out: Two Forms of Communication

Some new essential tools include websites, blogs, and monthly/quarterly newsletters. These can be used effectively once you understand there are two forms of communication to divide your efforts — let's call them initial contact and maintenance of existing support. To get your message across to a new potential consumer base, repeat the basic story you have designed from your earlier efforts. For those who have shown support in the past, you want to maintain their interest level and support by keeping them aware of ongoing efforts and giving them a sense of interaction. This can be as simple as an annual letter, but a quarterly or monthly e-mail, letter, or newsletter would be much better. Here are some tips:

Be regular. The level of contact will be your decision, but once you start, make every effort to be timely and consistent. Customers will notice when you skip contact.

Be engaging. Content should include your basic message, what is new around the farm, information on any new products, something seasonal linking your products to upcoming holidays, your hopes for the season, antics around the farm, arrival of turkey poults, turkey poults on pasture, and cooking tips and recipes that use your products.

Be year-round. Spread this contact around the year and include off-seasons such as winter. Much happens around the farm in winter—plans for next year, review of last year, vacation time for some of us — so keep your customer base informed and engaged.

Be grateful. Last, but importantly, be sure to thank your customers for your success. Let them know that their support of your farm is essential and appreciated.

Promote Yourself

There are some good promotions you can use to give your customers a sense of participation and to draw in new customers. You can use cook-outs, tastings at farmers' markets, farm tours, and even host harvest days where you offer customers a chance to come and help harvest the products. While not everyone will jump at the chance to process turkeys, you might be surprised at the things a portion of your customer base will enjoy: Many might enjoy helping load the turkeys. Others would gladly help build shelters or help dip the beaks of poults upon arrival.

◆ ◆ ◆

It takes good planning to successfully market the fruit of your turkey enterprise. But the necessary ingredients are really quite simple. First and foremost you must understand your market. Why do our customers buy turkey? What benefit are they seeking?

Here are some thoughts.

- In any given marketing effort you are presenting the consumer with features that distinguish the product, such as pasture raised, and perceived benefits, such as better flavor or nutrition.
- The product must be distinctive: look different, taste different, cook differently. In this way it will stand out from the competition. A good example is how heritage turkeys differ from commercial turkeys in appearance, flavor, and cooking.
- You are selling the story of your enterprise as well as badges such as organic, pasture-raised, heritage, humane, and so on.
- The section that most sells a turkey is the breast.
- While commercial turkey is available throughout the year in easily affordable forms, such as sliced turkey breasts for sandwiches, competing with industry for this "everyday" niche is not likely to succeed with the consumer — the price difference is too great. Only through connection with perceived additional benefits can we obtain premium prices for our products.
- Personal connections of consumers with their food will be your greatest road to success.
- Promotion can be as simple as including a card or brochure every time you sell a turkey, suggesting a turkey for coming events: birthdays, other national holidays, special dinners, parties. Include many recipe ideas — how to cook, but also what to do with leftovers.

12

Flock Health

SPECIAL THANKS *to Jeff May of Dawes Labs, Frank Reese of Good Shepherd Ranch, and Jim Adkins, Sustainable Poultry Specialist at the International Center for Poultry for their review and contributions to this chapter. Also special thanks to Leonard Mercia, who authored the original edition of* Storey's Guide to Raising Turkeys, *including the original version of this chapter. Mr. Mercia did such an excellent job that most of the disease descriptions in this chapter remain as originally written.*

A great hunter and a lover of nature, my grandfather had a very clear opinion on the responsibilities of raising any creature. He once sat me down, looked me in the eye and told me, "When you pen any creature, you become responsible for all of its care — its food, water, safety, and its health. It is a sin to neglect or give ill treatment to an animal when it is so reliant on your care."

Disease Prevention Starts with Good Health

The natural state of all turkeys is a healthy state. While disease-causing agents may be encountered in both natural and controlled environments, a healthy turkey with a healthy immune system is in the best position to remain healthy. To this end, we should attempt to utilize all the tools available to us to maintain this healthy state.

The basics of good health are a robust immune system, exercise, clean air, clean water, good foods, and a healthy environment. If we address each of these areas with good management, we can limit exposure to disease and ensure the turkeys have their best ability to resist or ward off diseases. So let's look at each in detail.

All livings things are imbued with immune systems — the world is full of life, including microscopic, and our turkeys must be able to protect themselves when other life forms seek to utilize them for food. The immune system is made up of a series of glands that utilize the circulatory system to distribute their benefits throughout a body. A healthy immune system is inherited through genetics. Faster rate of growth, as found in industrial/commercial turkey strains, is caused by an excessively active thyroid gland and this overactivity is linked with reduced immune function.

In a collaborative study by the American Livestock Breeds Conservancy and Virginia Polytechnic Institute & State University, conducted in 2004, several varieties of heritage turkey were compared to industrial turkeys for immune function and disease resistance. The researchers studied packed cell volume, a measure of red cells, which carry oxygen to the cells of the body, and total protein, which measures globulins and albumin. In both cases the heritage turkeys excelled over the faster growing industrial turkeys. Nonspecific T-cell stimulation and pan-lymphocyte stimulation were also studied — again, the heritage turkeys had superior responses.

Ascorbic acid production (vitamin C) was also studied since mammals (except humans) and birds are known to synthesize this for themselves. Ascorbic acid has been shown to reduce oxidative stress and protect from free-radical damage, modulate gene expression, and enhance immune function. Once again the heritage turkeys were superior.

Lastly, the turkeys were exposed to Hemorrhagic Enteritis Virus when six weeks old, and to *E. coli* 7 days later. Not surprisingly, the results were that all of the industrial turkeys died within 3 days, with most dying the first day, and most of the heritage turkeys surviving past 3 days and all the way through to the end of the study — in fact, none of the Black, Bourbon Red, or Slate turkeys died at all.

That immune function does vary from industrial to heritage turkey varieties should be a clear conclusion. And we should keep this in mind

WHAT IS NORMAL?

To understand signs and symptoms, you must know normal and abnormal turkey behavior. Normal Standard-bred turkeys are always active, curious, alert, running, jumping, flying, eating, drinking, fighting, and looking for things to do. Their feathers will be in good, shiny condition and their eyes clear and bright.

Abnormal is any change in the normal behavior or appearance, as simple as just wanting to stand in the corner.

Commercial turkeys suffer from the inability to perform normal turkey behavior; you may have to study the signs listed in order to recognize their condition or illness.

when choosing which variety best matches our system of production in order to have turkeys that will remain healthy. In the words of Jim Adkins, Sustainable Poultry Specialist for the International Center for Poultry, "After 25-plus years with Standard-bred turkeys, when I went to work in the commercial turkey industry, I had to LEARN a whole lot more about antibiotics, turkey medication, etc., because the immune systems of these birds are so low."

Exercise is perhaps the most overlooked necessity for living creatures such as turkeys. Exercise promotes movement of fluids and muscles, which aids in bringing cells hydration and energy. We can say that stagnation makes a body vulnerable to disease; the Chinese have said as much for centuries. Certainly, we see in humans the benefits of movement, where exercising the heart promotes good health and is, in fact, the treatment used to prevent most heart-related issues. Exercise leads not only to good muscle tone but to a good circulatory system. Any creature in a healthy state needs to exercise to remain in a healthy state.

Clean air is the next basic requirement for good health. Perhaps equally important as the circulatory system are the lungs and air sacs (the respiratory system). As some say, "the breath is life." Lung tissue and air sacs can be damaged by ammonia vapor escaping from manure and can become clogged, or even become a breeding ground for respiratory infecting agents, when exposed to large quantities of dust. All creatures need clean, fresh air to remain healthy.

Clean water is vital for good health. Bodies are 60–85 percent moisture. Moisture is lost through the skin, through the natural process of breathing, and through other bodily functions. In fact, dehydration is the primary agent of loss for poults during their first 3 days of life. Polluted water is a pathway for disease agents and can weaken a body by causing stress and reduced consumption. It is easy to see the importance of consuming clean water.

DISEASE PREVENTION GUIDELINES

SELECTION OF STOCK

- Select healthy fowl for breeders.
- Use pullorum-tested flocks.
- Hatch from hens rather than pullets (this assures longevity of the parents).
- Use males that come from lines with good liveability.

PROPER HOUSING

- Allow adequate floor space (this reduces cannibalism and spread of diseases by close contact and contaminated litter).
- Provide adequate ventilation, but avoid drafts. Avoid sudden changes in temperature. Drafts of extreme lower temperatures have been known to kill otherwise healthy birds.
- Use dropping pits or dropping boards to lessen the spread of diseases through droppings.
- Keep feeders and waterers clean. The more fecal material kept out of feed and water, the less chance of disease.
- Use litter and change it often.

SANITATION

- Clean feeders and drinkers daily, perches and nesting boxes weekly, and floors and the interior of the house periodically.
- Remove droppings and replace litter.
- Disinfect to kill any remaining disease-producing organisms. (Remember that most pathogenic bacteria and viruses soon lose their virulence and die when they are away from the body of the bird; chemicals are not always necessary.)

Good foods give the body energy to live and grow and are the blocks the body uses to build its cells and organs. Good foods provide protein, fats, fiber, minerals, and vitamins necessary for good health. Stale or spoiled feeds not only have reduced feed value, due to reduced values of vitamins, and so on, but become breeding grounds for molds that can stress a turkey's immune system. Turkeys thrive with a consistent diet of quality foods. Balancing what is fed is important as well. A turkey

PASTURE AND OUTDOOR SPACE

- 200 turkeys per acre is a good number.
- Rotate the birds within the pasture weekly.
- Do not return the birds to the same pasture for 3 years if possible (some disease germs and parasite eggs will live in a soil for a year or more).

FEED

- Proper feed not only protects turkeys against nutritional diseases but results in better vigor and greater resistance against infection by disease-producing organisms.

REMOVAL OF UNHEALTHY BIRDS

- Unhealthy birds in the flock should always be removed immediately, isolated, and either treated or culled to prevent infection of the rest of the flock.

PROTECTION FROM CARRIERS OF DISEASE

- Protect your flock from pigeons and wild birds — especially sparrows and starlings.
- Avoid feeding practices that attract flies.
- Young poults should be brooded a good distance from the adults.

QUARANTINE NEW OR RETURNING STOCK

- Allow a minimum of 14 days to ensure new birds do not show symptoms of disease — 30 days is preferable.
- If possible, bring new genetics in via eggs or poults from a tested flock.

Source: Jim Adkins, Sustainable Poultry Specialist, International Center for Poultry

cannot remain in excellent health when fed in excess of one particular feed, such as corn. A good diet is a balance of different grains, vitamins, minerals, and even protein and fat sources and is absolutely necessary to keep turkeys in good health. It should be noted that turkeys given liberty will tend to somewhat balance their own diets, as long as a variety of feeds can be scavenged.

The environment you choose to raise your turkeys in will have a lot to do with your success or failure. Think of the environment not only in broad terms, such as a confinement building or a pasture, but in finite terms of exactly what you will expose to your turkeys. Sun and rain are not problems for healthy turkeys as long as they have the ability to self-limit or adjust such exposure. One small wet area in a pasture is no problem as long as there are dry areas as well. Large amounts of dust or fecal matter are challenging whether encountered under a roof or on an open field. Wild birds and rodents have as much potential to expose turkeys to disease under a roof or on an open field — actually movement across a field may reduce the amount of exposure. Try to understand that turkeys must live and interact with their environment; as stewards, we are ahead of the game of good health when we understand the benefits and weaknesses of our system of production.

Ninety percent of the job of disease prevention is accomplished through managing our flocks to keep them healthy. By providing good management, good food, clean water, and fresh air with sanitary surroundings and limiting exposure to disease-causing agents, we will have fewer health challenges to address. A pound of prevention is worth a ton of cure.

AMMONIA AND LUNG DAMAGE

- 10 PPM causes damage to air sacs over several weeks.
- 25 PPM causes significant damage to air sacs and lungs in 1–2 weeks.
- 50 PPM causes damage to air sacs and lungs within 48 hours. This is the level where the human nose detects ammonia.
- 100 PPM causes mortalities.

CLOSE-UP ON THE ENVIRONMENT

Challenge	Effect	Solution
Dust	reduces respiratory function	allow plenty of air movement, add bedding material, move flock
Ammonia	reduces respiratory function, reduces growth, impedes immune function, death	allow plenty of fresh air, add more carbon bedding material, stir bedding material daily, move flock
Sun	death due to overexposure	provide shade
Rain	death due to chilling (at young age)	provide covered areas
Manure	causes disease (in large quantities)	reduce flock density per square foot/stocking rate per acre, remove manure, move flock
Swampy pasture	increased parasites, increased effects of manure exposure	move flock, limit exposure
Wild birds/ rodents	increased exposure to disease	trap or bait rodents, raise or cover feeders at night, reduce wasted feed to avoid attracting these pests, reduce flock stocking rate

A good biosecurity program is important for rearing turkeys or any other livestock. Biosecurity is an attitude, program, or management process that provides your birds with a rearing environment that is safe from many hazards and especially those related to disease. Biosecurity is important for all poultry flocks regardless of flock size.

The mechanics of biosecurity is to use practices that avoid exposure to contagious diseases. Vectors for disease can include wild birds, rodents, insects, and contact with contaminated clothing, boots, vehicles, and infected stock. Diseases are transmitted by close contact, but also via aerosol — such as by a sneeze or cough. Creating a biosecurity plan may sound laborious and unnecessary to a small producer; however, anyone with stock they value can benefit by taking a look at practices that can expose their flock to disease and adopting practices that limit the possibility of bringing in disease in the first place.

Considerations to protect our flocks will include: having dedicated footwear and clothing that is not worn off farm — the comingling of commercial poultry workers during lunchtime at restaurants, convenience stores, and gas stations is a major vector for disease; not sharing equipment or properly disinfecting equipment; controlling access of visitors to the flock; keeping vehicles that travel off farm away from poultry; preventing rodent and wild bird infestations; identifying and removing sick birds from the flock; using clean, noninfected water sources; and stepping up security during times of known disease risks.

Since having a plan — and thus being able to demonstrate that you follow a biosecurity program — can possibly give you a measure of protection when disease outbreaks, it is important that you demonstrate that you adhere to *all* of the practices of your plan. Do not choose practices you know you cannot follow. During the outbreak of foot-and-mouth disease in England in 2001, many herds and flocks in hot zones were spared the mandated culling when it could be demonstrated that they were of genetic importance and followed a biosecurity plan — they were quarantined and allowed to live.

Following a biosecurity plan does not guarantee disease prevention or that your flock will be spared in the case of a dangerous outbreak of disease. But following a biosecurity plan will greatly reduce the potential of disease transmission to your flock.

Disease Basics

Disease is a departure from a healthy state and includes any condition that impairs normal body functions. Infectious agents, including bacteria, viruses, fungi, and parasites, that cause disease in poultry can be introduced into a flock.

It is best to raise different types and ages of poultry separately. Keep young birds isolated from older birds and change shoes between visits to different age groups. Many poultry diseases have been transmitted on the soles of the shoes of good men. Do not allow unnecessary visitation to your flock and do not visit other poultry flocks if possible. Ideally, allow

Biosecurity

"Diseases should not be brought into the turkey community through the feed or on the feet of visitors or by winged visitors, the birds and pigeons."

— from *Turkey Management,* Marsden and Martin, 1945

ALBC MODEL PLAN

While writing a good biosecurity plan may seem difficult, and some of us will wonder where to start, there is an excellent model plan we can use. The American Livestock Breeds Conservancy offers a biosecurity plan that you can download for free from its website. The plan is customizable — you simply check off the practices you intend to adhere to and write in any additional practices you aim to follow. Visit http://www.albc-usa.org/documents/ALBCBiosecurityPlan.pdf for a copy of this plan.

no one who has poultry on your farm without proper sanitation, such as showering and change of clothes and shoes. A good rule is to never wear your on-farm boots or shoes off farm. If possible, do not visit your poultry the same day you have visited another poultry flock.

Preventive measures include a good sanitation program and a vaccination program designed to protect the flock from any diseases that may be prevalent in your area. A vaccination program should not be considered a substitute for good sanitation. Remember this: sanitation, isolation, eradication, vaccination, then medication.

If you purchase stock from a good, clean source, follow a sound sanitation program, use a good feeding program, and provide a comfortable growing environment, you have gone a long way toward keeping your flock healthy. However, losses do occasionally occur. Commercial flock owners, for example, expect a mortality rate somewhere between 8 and 15 percent. So if you lose one bird and the rest of the flock is eating and drinking and looking healthy, you don't need to get too worried about possible contagious disease as it is unlikely.

Many sick flocks do not show signs early. However, a disease in the flock is usually accompanied by several warning signs: (1) a drop (or dramatic increase) in feed or water consumption, (2) the appearance of sick or dead birds, and (3) a change in the birds' behavior and appearance. When it is apparent that a disease is present, seek the advice of a trained poultry diagnostician. Do not be afraid to find out the birds have some disease — most diseases are treatable. And remember, signs

RISK FACTORS FOR DISEASE TRANSMISSION

- Infected birds within a flock
- Newly acquired birds added to an existing flock, especially birds coming from shows or fairs and auctions
- Different species of birds reared together or in close quarters
- Different ages of the same species reared together or in close proximity
- Eggs (unsanitized) or poults from infected breeders
- Humans; hands, hair, feet or shoes, and clothes can harbor infectious agents
- Wild birds, rodents, flies, darkling beetles, other insects, and parasites
- Contaminated feed, water, or air
- Contaminated vaccines and medications
- Contaminated equipment brought onto the farm, such as trucks, tractors, used coops, and egg flats
- Vaccines that are so potent that they cause the disease rather than prevent it

that appear today usually are the results of problems 2 to 3 weeks ago. Do not just use drugs or antibiotics indiscriminately; this may be of little value and only result in a waste of money.

It is a good idea to build a relationship with someone who has contacts with diagnostic laboratories. It is difficult to get good, useful testing unless you know what you need to test for and ask for that specific test. Labs do lab work, not veterinarian work; your rep will help you locate the best lab for your problem. These representatives can be found through your suppliers of feed, baby birds, or medications, and through county Extension agents. If there are no local diagnosticians, you may submit sample birds to a state diagnostic laboratory. The sample should include two or more sick, or recently dead, birds. Preserve dead specimens by keeping them cool to prevent decomposition. Early diagnosis and fast treatment are always recommended as the quickest ways to solve poultry disease problems. The addresses of the state diagnostic laboratories can be found in the back of this book. Alternatively, you can contact your local county Cooperative Extension Service by looking in the phone book under County Government. Ask for a poultry or livestock agent. This person can help you contact the state diagnostic laboratory and can address many management issues.

Turkey Diseases

Several diseases and parasites may affect turkeys. Only the more common ones are described here, and these are not discussed in great detail. For more in-depth information, you can consult many excellent texts on poultry diseases (see Resources). Also, discuss poultry disease issues with some of your rural veterinarians. If you can find one with an interest in birds, especially poultry, he can be of invaluable assistance. This would be especially true if the veterinarian is consulted in conjunction with a state diagnostic laboratory.

Recognizing a disease problem at the onset, diagnosing it before it becomes widespread in the flock, and getting treatment started early can greatly reduce the possible losses due to mortality, morbidity, and diminished overall performance.

Aspergillosis (Brooder Pneumonia)

Aspergillosis is an environmental disease that is prevented through good management. The symptoms include the following:

- The birds stop eating.
- Breathing may be rapid.
- The birds may gasp and have labored breathing.
- Eyes may be inflamed.
- Eyelids may swell and stick together.

This disease is caused by a fungus that is inhaled by the birds and usually comes from moldy litter or feed. On the postmortem examination, yellow-green nodules may be found on the lungs and in the trachea, bronchi, and viscera.

PRECAUTIONS FOR DRUG AND PESTICIDE USE

To maintain a healthy flock and to obtain optimum production, it is sometimes necessary to administer drugs or pesticides. Use all drugs and pesticides according to the label directions, and use them with caution. Never use a pesticide that is not registered for use on poultry.

For treatment of any disease, but especially if drug administration is necessary, you should involve a veterinarian. Although it may take effort to locate a local veterinarian who treats birds, it will be worth the effort. Also, be sure to utilize your state livestock diagnostic services. For more information, contact a poultry or livestock Extension agent at your local Cooperative Extension Service. Look in the phone book under Local or County Government.

TREATMENT AND PREVENTION

If aspergillosis has been diagnosed by laboratory testing, thoroughly clean all buildings and equipment, removing all fecal material and plant matter, such as straw. Replace bedding with pine shavings. Infected birds will not get better but generally are not contagious to others.

This is a disease of young birds if it has originated at the hatchery. If so, inform the hatchery so they may clean up also.

Spread of infection may be prevented by culling the sick birds, thoroughly cleaning and disinfecting the house and equipment, and carefully removing moldy litter or feed from the building.

Avian Influenza

More than one hundred influenza-virus strains may infect birds, the majority affecting ducks and turkeys. The disease can be mild or acute. The mild form produces listlessness, respiratory distress, and diarrhea; the acute form causes air sacculitis and sinusitis with cheesy exudates. Large drops in egg production can occur.

The best prevention is isolation from other flocks. Avian influenza seems to become a problem when husbandry and sanitation are below par. Keep wild birds, especially migratory waterfowl, away from turkey flocks. The risks are even greater if you keep waterfowl that attract wild birds during the spring and fall migration periods.

If avian influenza is known to be a problem in an area, keep in mind that the organism can be transmitted in many ways from one farm to another. For example, it can be transmitted on clothes, equipment, egg boxes, and poultry crates.

TREATMENT AND PREVENTION

Blood-test for the disease and eradicate known infected flocks. Remember to be kind to your neighbor by not harboring disease.

Blackhead (Histomoniasis)

Blackhead is caused by the protozoan parasite *Histomonas meleagridis*. The signs of this disorder include drooping heads, dark heads, and brownish, foamy droppings. On necropsy, inflammation of the intestine and ulcers on the liver may be seen. The term *blackhead* is somewhat misleading because that sign may or may not be present.

This disease affects turkeys of all ages. It can also affect chickens; however, the disease tends to cause less mortality in those birds. Since chickens may act as an intermediate host for the organism that causes blackhead, it is recommended that they not be kept in the same house and never be intermingled with turkeys. Ideally, they should not be kept on the same farm.

CLOSE-UP ON BLACKHEAD

You may first identify blackhead due to a bird's sulfur-colored diarrhea (and later its spotted liver). Kill infected birds or move them to sun porches for treatment. If you move the flock to fresh pasture once or more per month, blackhead seldom causes serious losses.

A milk flush can be used to treat blackhead or coccidiosis. It consists of feeding mash with 25 percent dried milk for 1 day to flush out the bird's digestive tract.

Cecal worm eggs can harbor the organism that causes blackhead for long periods. When picked up by the turkeys, it infects the intestines and liver. Both chickens and turkeys can host the cecal worm. Blackhead can live in the soil via the worm for many years. Mortality with this disease may reach 50 percent if treatment is not started and the infection checked immediately.

TREATMENT AND PREVENTION

Incidence and severity of the disease depend on the management and sanitation programs used. If you do not have blackhead today, prevent it from getting to your farm; it takes years to be rid of it. Several measures are helpful in preventing blackhead:

- Follow good sanitation practices in the brooding facilities.
- Rotate the range areas.
- Segregate young birds from old birds.
- Separate turkeys from chicken flocks.

Coccidiosis

This common poultry disease is caused by *Coccidia*, a group of protozoan parasites. The birds become exposed by picking up sporulated oocysts in fecal matter and litter. All flocks grown on litter or range are vulnerable. Birds raised on elevated wire or slats are not exposed to droppings and normally don't contract coccidiosis, but even they are at risk if feces are retained in the pen or contaminate the feed or water.

Coccidia are host specific: that is, the coccidia that affect turkeys do not affect chickens. Different species of the parasite affect different parts of the digestive tract. Six species are known to infect turkeys, but only three of them are commonly troublesome. If left unchecked, the disease can be fatal.

Every bird will be infected with coccidiosis. The disease symptoms are from an overwhelming amount of the organism in a less-than-healthy bird. Very young birds need to build immunity to the organism and are most susceptible to the disease.

You should suspect coccidiosis in your flock if you notice the following:

- Ruffled feathers
- Bloody diarrhea

- Head drawn back into the shoulders and the appearance of being chilled (birds having this appearance are sometimes called *unthrifty*)

Necropsy findings may include lesions and hemorrhages in various parts of the intestine, depending on the particular species of parasite.

TREATMENT AND PREVENTION

Coccidiosis may be prevented or controlled by feeding anticoccidials at low levels in the starter feed. Apple cider vinegar can be offered in the drinking water as a natural anticoccidial — the increase in acidity discourages the coccidia. Also, keeping young birds on wire, so that they do not come in contact with their feces, will keep the parasites from overwhelming the youngsters. A caution here, however: Lack of contact with the coccidia organism at an early age can result in increased chance of mortality if exposed as an adult.

Erysipelas

Erysipelas, which means red skin, is caused by the bacterium *Erysipelothrix insidiosa*. Swine, sheep, humans, and other species are also susceptible to the disease. The signs of erysipelas are swollen snoods, bluish purple areas on the skin, congestion of the liver and spleen, listlessness, swollen joints, and yellow-green diarrhea.

Erysipelas is primarily a disease of toms because the organism readily enters through wounds caused by fighting. Since the snood is frequently injured when toms fight, this is a common site for erysipelas infection. For this reason, some commercial producers have their turkeys' snoods removed at the hatchery or on the farm upon arrival. Erysipelas is a soil-borne disease, and contaminated premises are the primary source of infection.

TREATMENT AND PREVENTION

The disease responds well to penicillin, and tetracycline is also effective. However, consult a veterinarian for treatment.

Control requires good management and sanitation. Vaccination is recommended for areas in which the disease is common. If this disease is suspected, use care. Wear gloves when performing a necropsy on a diseased bird.

Fowl Cholera

Caused by the bacterium *Pasteurella multocida*, this disease is highly infectious and affects all domestic birds, including turkeys. The birds become sick rapidly and may die suddenly without showing signs. When signs do appear, they include listlessness; fever; excessive consumption of water; diarrhea; swelling of the head, face, and sinuses in the chronic form; red spots or hemorrhages on the surface of the heart, lungs, or intestines, or in the fatty tissues on postmortem examination; and swollen liver (that is, the liver has a cooked appearance with white spots).

TREATMENT AND PREVENTION

Chlortetracycline can be used.

Good management practices are essential to prevention. Sanitary conditions in the poultry house, range rotation, and proper disposal of dead birds help prevent cholera. In problem areas, vaccines can be used and are recommended. Do not let any pregnant animals eat the dead carcasses.

Fowl Pox

Found in many areas of the United States, fowl pox is caused by a virus and spreads through contact with infected birds or by such vectors as mosquitoes and other biting insects or wild birds.

There are two forms of fowl pox — the dry, or skin, type and the wet, or throat, type.

Birds with fowl pox have a poor appetite and look sick. The wet pox causes difficult breathing; nasal or eye discharge; and yellowish, soft cankers of the mouth and tongue. The dry pox causes small, grayish white lumps on the face, which eventually turn dark brown and become scabs. On postmortem examination, cankers may be found in the membranes of the mouth, throat, and windpipe. There may be occasional lung involvement or cloudy air sacs.

TREATMENT AND PREVENTION

Although the disease has no treatment, antibiotics may help to reduce the stress associated with it. The only means of control is by vaccination, which is recommended in areas where fowl pox is a problem or when infection has been identified in your flock.

SANITATION IS KEY TO GOOD HEALTH

Practically all turkey diseases are spread through infected droppings. Controlling turkey disease, therefore, consists almost entirely of keeping the flock away from its own droppings.

If turkeys must pass repeatedly through a gate or lane, this area may be contaminated with droppings. Birds may congregate if shade is sparse. To prevent disease in late fall, and other periods of cold rain:

- Move your birds more often
- Watch for and fill water puddles
- Move feeders and fountains a few feet every day to prevent contact with the accumulated manure
- Remove dead birds
- Isolate sick birds
- Avoid wet mashes, as they attract swarms of flies, or offer feed in small amounts

Mycoplasma-Related Diseases

Mycoplasma bacteria may cause several types of disease conditions in turkeys and other bird species. There are several strains of *Mycoplasma* bacteria, but those of primary concern are *M. gallisepticum, M. synoviae, M. iowae,* and *M. meleagridis* (see below). Outbreaks of disease caused by these organisms result in a variety of symptoms and bring about poor growth rates and egg production, along with possible flock morbidity and mortality.

Mycoplasma organisms are extremely small compared to other bacteria and do not have a rigid cell wall. They can survive for up to several days outside the bird on feathers, clothes, and hair, for example.

TREATMENT AND PREVENTION

Test your flock and if it is infected with any of the mycoplasmas, work with your diagnostic representative to eradicate the disease. It will be difficult but very rewarding. Mycoplasma flocks have many other disease problems that cause high death losses.

Once a flock is infected with this disease, the best course of action is to depopulate the farm and then clean and disinfect everything. Have a down time of at least 2 weeks and then restart production.

Infectious Sinusitis

Infectious sinusitis is a disease of turkeys caused by *Mycoplasma gallisepticum* — the same organisms that cause chronic respiratory disease in chickens. The disease is also found in pigeons, quail, pheasants, ducks, and geese. These bacteria are transmitted through the egg from carrier hens. Stress is thought to lower the poult's resistance to the disease (this tends to be true for most diseases).

Affected birds show nasal discharge, coughing, difficulty breathing, foamy secretions in the eyes, swollen sinuses, decreased feed consumption, and weight loss. Air sac infection may be in evidence on postmortem examination.

TREATMENT AND PREVENTION

Antibiotics in the feed or water are useful to help control mycoplasma infections. Individual treatment with injectable penicillin and streptomycin in the sinuses can also be useful. Obtaining poults from *Mycoplasma*-free breeding stock is the most important aspect of disease control.

Infectious Synovitis

An infectious disease of turkeys caused by *Mycoplasma synoviae*, synovitis was first identified as a cause of infections of the joints, but more recently it has been shown to cause respiratory disease as well. This disorder can affect birds of all ages, causing lameness, reluctance to move, swollen joints and foot pads, weight loss, and breast blisters. Some flocks have respiratory symptoms. Greenish diarrhea occurs in dying birds.

The most common means of transmission of synovitis is through infected breeders. Poor sanitation and management practices also contribute to the problem.

Postmortem findings include swelling of the joints; presence of a yellow exudate, especially in the hock, wing, and foot joints; possible signs of dehydration; enlarged liver and spleen; and air sacs filled with liquid exudate. Aside from findings on necropsy, respiratory involvement is not easy to spot.

TREATMENT AND PREVENTION

Antibiotics yield some results, and they should be given by injection or in the drinking water. Some producers prefer to give antibiotics by both methods simultaneously, though this can result in difficulty in managing dosage. Always obtain poults from *Mycoplasma*-free breeders.

Mycoplasma iowae Infection

Shown to be responsible for reduced hatchability in turkeys, *Mycoplasma iowae* is transmitted through the egg from the breeder hen like the other types of mycoplasma. It can be lethal to turkey embryos.

TREATMENT AND PREVENTION

The disease is best prevented by obtaining poults from *M. iowae*–free breeder flocks.

Mycoplasma meleagridis Infection

Like the other mycoplasma disorders, *Mycoplasma meleagridis* is an infectious disease of turkeys that a breeder hen transmits to the egg. The main sign is air sacculitis (inflammation of the air sacs). Even though this type of infection is thought to be specific to turkeys, it may occur in peafowl, quail, and pigeons. Obtain poults from *M. meleagridis*–free stock.

Newcastle Disease

Acute and highly contagious, Newcastle disease is a respiratory disorder that is caused by a virus and is found in chickens, turkeys, and other species of poultry. It causes high mortality in young flocks. In breeder flocks, egg production frequently drops to zero. Newcastle spreads rapidly through the flock.

Signs of the disease are gasping, coughing, hoarse chirping, increased water consumption, loss of appetite, huddling, partial or complete paralysis of the legs and wings, and holding of the head between the legs or on the back with neck twisted. Postmortem examination may reveal congestion and hemorrhages in the gizzard, intestine, and proventriculus; cloudy air sacs may also be noted.

The disease is transmitted in many ways: it can be tracked in by people or brought in by birds from another site, dirty equipment, feed bags, or wild birds.

TREATMENT AND PREVENTION

There is no effective treatment, though antibiotics are normally given to limit secondary infections.

Vaccination is recommended in most areas of the country and can be administered to an individual bird or on a mass basis. On an individual basis, the birds can be vaccinated intranasally, ocularly, or in the wing web. (The wing web is the thin layer of skin at the forward edge of the wing between the proximal or shoulder end of the humerus and the tip of the wing.) On a mass basis, the vaccine can be given to the birds in drinking water or in the form of a mist or spray. Follow the manufacturer's recommendations when using these products, and conform to the vaccination program that is recommended for your area.

Omphalitis

Omphalitis is caused by a bacterial infection of the navel and occurs when the navel doesn't close properly after hatching. It can also be caused by poor sanitation in the incubator or hatchery, chilling, or overheating.

Signs of omphalitis may include weakness; unthriftiness; huddling; an enlarged, soft, mushy abdomen; and an infected navel surrounded by a bluish black area. Mortality may be high for the first 4 or 5 days of life.

TREATMENT AND PREVENTION

There is no treatment for the disease. Most of the affected poults die within the first few days, and no medication is needed for the survivors. Report this problem to your poult suppliers; they need to make changes to prevent reoccurrences of the problem.

Salmonella-Related Diseases

More than two thousand species of the genus *Salmonella* have been identified. Although quite a number of these species can affect chicken and turkeys under certain conditions, very few are a serious threat to the poultry industry. Of greatest concern are *S. pullorum*, *S. gallinarium*, *S. arizonae*, and paratyphoid infection. Paratyphoid is caused by many species of *Salmonella* and can infect four-legged animals as well as poultry.

Pullorum

An infectious disease of chickens, turkeys, and some other species, *Salmonella pullorum* (sometimes called white diarrhea) is found all over the world. The National Poultry Improvement Plan was organized in 1935 by the U.S. Department of Agriculture to eradicate pullorum as well as fowl typhoid. The disease causes high mortality, which most often occurs at 5 to 7 days of age.

S. pullorum is transmitted from the hen to the poult mainly through the egg. After transmission, it spreads rapidly through the down of poults located in incubators and hatchers. Birds with pullorum appear droopy, huddle together, act chilled, and may have diarrhea and pasting of the vent. Examination of diseased chicks reveals dead tissue in the heart, liver, lungs, and other organs and an unabsorbed yolk sac. The heart muscle may be enlarged and have grayish white nodules. The liver may also be enlarged, appear yellowish green, and be coated with exudate.

TREATMENT AND PREVENTION

One of several types of blood tests can help establish a positive diagnosis. Flocks that have pullorum should be depopulated or destroyed immediately and definitely not kept as replacements or for breeding purposes. Buy poults from pullorum-free hatcheries only.

Fowl Typhoid

Caused by the bacterium *Salmonella gallinarum*, fowl typhoid affects chickens, turkeys, and other species of birds and may be present wherever poultry is grown.

Affected birds may look ruffled, droopy, and unthrifty and have a loss of appetite, increased thirst, and yellowish-green diarrhea. Postmortem examination may show a mahogany-colored liver, an enlarged spleen, and pinpoint necrosis in the liver and other organs.

TREATMENT AND PREVENTION

Typhoid is prevented in the same manner as pullorum: buy typhoid-free poults. Flocks that are positive for fowl typhoid should be destroyed.

Arizona

Salmonella arizonae causes an infectious disease that can affect chickens but most commonly strikes turkeys. This disease, which is also called paracolon infection, has both acute and chronic forms. Many serotypes of the disorder occur in mammals, birds, and reptiles. Mortality usually occurs in the first 3 to 4 weeks of life. There can be high morbidity (that is, sickness) without high mortality.

The disease has no distinct signs; however, unthriftiness and blindness may occur. Infections of the intestinal tract, peritonitis, and mottled and enlarged livers may appear on necropsy.

Diagnosis is based on laboratory isolation of the organism. Various drugs are used to minimize mortality from this form of salmonella in poults. The disease is most commonly transmitted by hen to egg to poult, but it can also be spread by direct contact with infected birds, rodents, and contaminated premises.

TREATMENT AND PREVENTION

Blood testing of breeders and ensuring proper sanitation of the hatchery and other environments are important for prevention. Good rodent control is imperative for control of this and other salmonelloses.

Paratyphoid

An infectious disease of turkeys and some other birds and animals, paratyphoid is caused by one or more of the *Salmonella* bacteria other than those discussed in the preceding sections. Transmission may be from the hen through the egg to the chick. The organism is also found in fecal matter of infected birds.

The disease primarily infects young birds but may also affect older birds. In young birds, mortality can run as high as 100 percent.

Some birds may die of paratyphoid without showing signs; however, you may notice weakness, loss of appetite, diarrhea, and pasted vents. Birds may appear chilled and huddle together for warmth. Older birds lose weight, are weak, and have diarrhea.

On necropsy, birds that have died of paratyphoid reveal unabsorbed yolk sacs, small white areas on the liver, inflammation of the intestinal tract, congestion of the lungs, and enlarged livers. Older birds may have white areas on the liver, but most typically show no lesions.

TREATMENT AND PREVENTION

Some antibiotics may reduce losses, prevent secondary invading organisms, and increase the bird's appetite. The disease can be controlled through sanitation and isolation of the flock from sources of infection, such as wild birds, birds from other flocks, rodents, and contaminated feed and equipment.

Turkey Coronavirus (Bluecomb)

Turkey coronavirus is a highly contagious disease of turkeys of all ages. Signs include depression, subnormal body temperature, diarrhea, loss of appetite, weight loss, poor growth, poor feed conversion, watery feces, dehydration, and prostration.

Some flocks of turkeys with coronavirus seem to be healthy and show few signs of the disease. However, flocks that test positive for this virus

REDUCING THE CHANCE OF REINFECTION WITH CORONAVIRUS

If a flock tests positive for coronavirus, do the following:

1. Depopulate the farm.
2. Clean and disinfect everything.
3. Let the facilities sit empty for 4 to 6 weeks.

Be careful not to spread litter from infected flocks around other turkey or poultry flocks. Some turkey producers have observed that letting litter sit undisturbed for 2 weeks before removal from the turkey house cuts down on transmission.

usually do not perform as well (with respect to growth and feed conversion) as flocks that test negative, even when they do not show signs. Mortality can be very low or extremely high in poults but is usually low in older birds. Coronavirus is spread by direct or indirect contact with infected birds or contaminated premises.

TREATMENT AND PREVENTION

Some antibiotics may help in cases where there is risk of high mortality with secondary bacterial problems. However, good husbandry and management and strict adherence to good biosecurity practices are the best prevention for coronavirus.

Turkey Parasites

Several parasites — both internal and external — can affect poultry, but relatively few of them are of major importance.

Internal Parasites

Some internal parasites may cause setbacks in weight gain and a loss of egg production in laying birds; severe infestation can cause death. Some intestinal parasites harbor other disease organisms that may be harmful to turkeys. Good management and the type of management system used are key in the control of internal parasite infestations.

Interestingly, the U.S. Army put out a poultry handbook for domestic use during World War II. In this book, Black Walnuts with green or bruised husks were used to rid poultry of internal parasites. Three to five walnuts per gallon of water were allowed to soak overnight in a bucket. In the morning, the "medicated" water could be mixed to its proper dosage to average one walnut per gallon, and this was the only water given to the birds. Treatment was to repeat in 2 to 5 days.

Large Roundworm

Light infestations of roundworms are probably not a cause for concern. However, when the worms become numerous, birds can become unthrifty and feed conversion and weight gain suffer. These worms may also reduce egg production. By themselves, the worms rarely kill their host, but they can cause fatalities if they occur along with other diseases.

The large roundworm is 1 to 3 inches (3.8–7.5 cm) long. It is found in the upper to middle portion of the small intestine. If the birds are heavily infested, the worms may extend for the full length of the small intestine. Piperazine and other wormer compounds are used to treat birds with roundworms and can be given in the water, the feed, or a capsule.

Clean, dry litter aids in the control of roundworm infestations in the growing houses. Where turkeys are free-ranged, regular rotation of the range area has been found to be quite effective in controlling worm infestations.

Cecal Worms

Cecal worms are very small and by themselves are not injurious. However, they are significant because they act as carriers for the organism that causes blackhead (see page 261).

Gapeworm

Gapeworms attack the bronchi and trachea and can cause pneumonia, gasping for breath, and even suffocation. Small turkeys open their mouths with a gaping movement and may have bloody saliva. Mortality may be high among young infected birds. Gapeworms can quickly build up drug resistance; therefore, treatment should be tailored for each farm

CONTROLLING INTERNAL PARASITES

Effective control of internal parasites depends primarily on a program of cleanliness and sanitation. Parasite eggs can remain viable in the soil for more than a year. This means that it's important to rotate poultry runs or yards. Preferably, poultry ranges should be used for 1 year and left idle for 3 before they are used again. Poultry yards and runs should be located in well-drained areas and be kept as clean as possible. Cultivating and seeding down these areas helps prevent the birds from picking up parasite eggs.

or flock. The earthworm is an intermediate host. The gapeworm is fork-shaped and red in color.

Tapeworm

There are several species of tapeworms, varying in size from microscopic to 6 to 7 inches (15–17.5 cm) in length. They are flat, white, and segmented and inhabit the small intestine. They cause weight loss and lowered egg production. Tapeworms need intermediate hosts like worms, snails, or beetles to complete part of their life cycle. Turkeys get tapeworms by eating the infected worms, snails, or beetles.

External Parasites

Although there are many external parasites of poultry, few are of major importance. However, certain external parasites, especially when present in large numbers, can cause loss in weight or loss of egg production, as well as decreased growth rates and feed efficiency. Only the more important external parasites are discussed here. Birds should be handled and closely observed on a regular basis to catch external parasite infestations early.

Treatment for external parasites usually involves dusting them with powdered pesticides or spraying them with liquid versions of the same. For treatment to be successful, it is important to treat the parts of the body that parasites will tend to favor, such as the breast, between the legs, around the vent area (but never allow any pesticide to enter the vent or risk toxic shock and loss of fertility or life), on the back at the base of the tail, on the back of the neck at the hackle area, and under both wings. Treatment should be repeated once per week for a total of three treatments in order to be sure to kill the adult parasites as well as any that hatch from eggs.

Lice

Lice are chewing and biting insects that cause birds considerable grief. With severe infestations, growth and feed efficiency suffer; lice can also affect egg production. They irritate the skin and result in scab formation.

Lice spend their entire lives on the birds and die within a few hours if separated from the host. The eggs (nits) are laid on the feathers,

TREATING AND PREVENTING LICE

Insecticides that may be used to treat lice are carbaryl, malathion, coumaphos, and pyrethrins or permethrin. Treat the birds according to the directions on the label, and examine them frequently for signs of reinfestation. It is usually the case that to get rid of lice, it will be necessary to treat the flock once per week for 3 consecutive weeks — thus killing the newly hatched lice before they may reproduce.

Wood ashes may be sprinkled liberally in areas where turkeys dustbathe. This is an excellent preventative, and the birds self-treat. Be advised, however, that the turkeys will consume some of the charcoal bits and this, while being a benefit and pulling out toxins from the body, will also cause the manure to be black for a few days.

where they are held with a gluelike substance. The eggs hatch within a few days up to 2 weeks after being laid. Lice live on the scale of the skin and feathers. Several types attack poultry. Lice can be gray or yellow; it's difficult to distinguish between the colors. The body louse, one of the most common poultry lice, usually affects older birds. The lice and their eggs are seen on the fluff, the breast, under the wings, and on the back.

Mites

A number of species of mites are capable of influencing flock performance. Some live on the birds; others spend more time off the birds. Mites' mouths are adapted either to chew or to pierce. They live on blood, tissue, or feathers, depending on the type. Generally, mites cause irritation and affect growth and egg production. In the case of severe infestations of certain mites, they can cause morbidity, or even mortality, in the flocks.

Monitor for mites. Mites can be particularly troublesome, in respect to both their effect on turkeys and the effort needed to eradicate them. Keep a close and watchful eye on your birds to catch mite infestations early so that treatment will be effective. In general, a quick way to test your flock is to check the tom turkeys — if they have mites or lice, then it is likely the hens will too.

The Northern Fowl Mite. Northern fowl mites are a reddish, dark brown. These mites are found around the vent, tail, and breast and live on the birds at all times. They attach to feathers and suck blood, causing anemia, weight loss, and reduced egg production.

Materials recommended for treatment of northern fowl mites are carbaryl, malathion, and pyrethrins (permethrins). Use according to the manufacturer's directions. It is usually the case that to rid the flock of mites, it will be necessary to treat the flock once per week for 3 consecutive weeks — thus killing the newly hatched mites before they may reproduce.

The Chicken Mite. Also known as red mites, chicken mites feed at night and are not found on the birds during the day. During daylight hours, they may be seen on the underside of roosts, in cracks in the wall, or in seams of the roosts. Other signs are salt-and-pepper–like trails under roost perches and clumps of manure. Red mites are bloodsuckers and cause irritation, weight loss, reduced egg production, and anemia.

Treat the chicken mite with the same insecticides as those used for the northern fowl mite. Roosts should be treated, especially cracks and crevices. An old method was to paint the roosts with kerosene in the morning, to allow the fumes to dissipate while killing the sleeping mites. Linseed oil works well for a roost "paint" as do liquid forms of the insecticides. Both kerosene and linseed oil have the drawback of being flammable, so use caution when using either of these methods.

CHEMICAL SAFETY MEASURES

Keep drugs and pesticides in their original containers in a locked storage area, well out of reach of children and animals. Avoid inhaling sprays or dusts, wear protective clothing, use recommended equipment, and be safe.

Disease Prevention

There is no substitute for good management. Prevention is worth a lot more than treatment. Drugs or pesticides are not intended as substitutes — they work best when combined with good sanitation and sound management practices. Early diagnosis and treatment of a disease or parasite problem is important.

Probiotics

An alternative to antibiotics is a class of feed additives called probiotics. These are fermentation products that contain either live cultures of beneficial bacteria or by-products of fermentation, such as mannan oligosaccharides (MOS). The term *saccharide* refers to sugars, and MOS products are complex sugars that are not digested by the animal. Instead, they attach to pathogenic bacteria and prevent them from adhering to the gut wall. The animal's digestive systems can then remove the bound bacteria from its system. Probiotics are beginning to gain favor with poultry producers. However, if you want to avoid antibiotics but wish to provide some protection for your turkeys, probiotics may be worth a try.

One source of live-culture probiotics is from raw milk and raw milk by-products. During the early 1900s, and before, it was common practice to use this food source to be fed as either liquid buttermilk, or the raw milk or whey was used to make a wet mash — given only in the quantity that could be consumed in 15 minutes or less. It seems this time-honored practice has a great deal of merit.

Drug Withdrawal Periods and Tolerance

Federal agencies have established withdrawal periods and tolerance levels for various agents used in poultry production. For example, the U.S. Food and Drug Administration (FDA) insists that certain drugs be withdrawn a specified number of days prior to slaughter, and some pesticides cannot be used within a certain number of days. These withdrawal periods vary from 1 or 2 days to several days and are subject to change. The FDA also establishes maximum amounts of residues for certain chemicals. Some insecticides can be used around poultry but not directly on the birds, on the eggs, or in the nests. Frequency of use may also be restricted for some agents.

Because withdrawal periods and tolerances and accepted forms of treatment do change, specific precautions for various agents are not addressed here. The point is that drugs and insecticides must be used discriminately. Follow all precautions on the label. If used improperly, drugs and insecticides can be injurious to humans, animals, and plants.

Nutritional Deficiencies

A number of poultry disorders may be caused by nutritional deficiencies or imbalances. With today's well-formulated diets, nutritional problems occur infrequently, so a thorough discussion of nutritional deficiencies will not be undertaken here. However, to underscore the importance of good nutrition, a few of the more common nutritional problems will be mentioned.

Rickets

Rickets is caused by a deficiency of vitamin D_3, phosphorus, or calcium or by a calcium-phosphorus imbalance. It may occur in birds on range or when grains are used along with complete feeds or protein concentrates because the birds may not get enough calcium. Provide oyster shells to these birds.

Birds with rickets exhibit weakness; stiff, swollen joints; soft beaks; soft leg bones; and enlarged ribs.

Perosis

Perosis in turkeys, sometimes called *slipped tendon*, is a leg problem typically caused by a deficiency of choline in the diet. Heredity may also be a factor. Other dietary factors include biotin, folic acid, manganese, zinc, and possibly pyridoxine. Swollen hocks can also be caused by deficiencies of niacin and vitamin E. In perosis, the large tendon of the leg at the rear of the hock slips to one side, resulting in a twisted leg. If permanently crippled, the bird should be killed. Most turkeys with perosis respond to early use of additional manganese in the feed.

DO I KNOW YOU?

Turkeys can recognize a new turkey in their midst. A flock of turkeys will almost always attack the newcomer, resulting in wounds and, sometimes, a fatality. It is better to integrate new turkeys by placing them in an adjoining pen so that the flock may see them for a week or more.

Miscellaneous Problems

Some problems do not easily fit into the aforementioned categories but do need to be described. They will probably be easily observed when they occur in your flock.

Good management will prevent some of the miscellaneous problems. However, on occasion a few birds in a flock develop abnormalities and have to be removed from the flock. Leg problems, for example, may make it difficult for the affected birds to get to feeders and waterers. Sometimes these birds can be segregated from the flock and nursed back to health, but frequently they have to be sacrificed in a humane manner. This type of mortality, plus normal mortality, requires a sanitary bird-disposal program. Rodents are also a common problem on most poultry farms, making a rodent-control program necessary.

Pendulous Crop

In the normal position, the crop is in the wishbone cavity and is attached to the side and back of the neck. If for some reason the connective tissues that hold the crop in place weaken, the crop drops. If the crop gets too far out of its normal position, feed cannot pass from the crop to the gizzard and the bird actually starves with the crop full of feed. Young birds with a mild pendulous crop condition may recover. Seriously affected birds seldom recover, and treatment is ineffective.

Leg Weaknesses

Leg weaknesses other than perosis may be caused by vitamin deficiencies or by such diseases as infectious synovitis. In day-old poults, a condition resembling perosis is called *spraddle legs*. This disorder may be due to a genetic factor, faulty incubation, or a deficiency in the diet of breeding stock. Smooth, slippery surfaces in hatching trays, shipping boxes, or under brooders that cause poor footing may also result in spraddle legs. Place young poults on wire, paper with a rough surface (if paper is used), or litter to avoid spraddle legs. Another leg problem, crooked toes, may be hereditary or due to faulty brooding conditions.

Euthanasia

Anyone who decides to raise or keep livestock should consider how to humanely kill sick and crippled animals that need to be removed from the group. If you are raising turkeys for slaughter, you might take euthanasia in stride. However, if you are keeping turkeys as a hobby flock in which the birds reach a near-pet status, the situation may be quite different and even traumatic. Even in hobby flocks, there are times when birds should be humanely killed. For owners of such flocks or people who view their birds as pets, the use of a veterinarian might be best advised. However, producers can perform euthanasia for their birds when needed. The options for euthanasia include blunt trauma, cervical dislocation, carbon dioxide overdose, and drugs.

Blunt Trauma

In this context, blunt trauma is the striking of the head with a heavy, blunt instrument, 6 to 8 inches (15–20 cm) in length. This may sound crude and objectionable to some; however, it can be very effective and humane if done properly. Turkeys do not have thick skulls, and a heavy blow with an appropriate object kills the bird immediately. The turkey should be restrained so that the blow is properly placed to kill the bird rather than to cause injury. Blunt trauma is usually considered for older animals but is effective for young birds as well.

Cervical Dislocation

This method of dispatch involves dislocation of the neck vertebrae from the cranium. However, it is important to realize that cervical dislocation includes the separation of the spinal cord and carotid arteries. When done properly, the bird is killed instantly.

WARNING!

If CO_2 overdose is your preferred method of dispatch, note that a box built for a large bird would also be large enough for a small child. An overdose of CO_2 kills a person as readily as it does a turkey. For this reason, construct small CO_2 boxes for use on small groups of young birds. Keep CO_2 tanks locked in a secure location when not in use.

Carbon Dioxide (CO_2) Overdose

Another method of humane death, the carbon dioxide chamber can be simply a box with a lid on the side or top. To construct the chamber and dispatch a bird:

1. Drill a hole in the box for a hose from a CO_2 tank.
2. Place the bird in the box.
3. Close the lid.
4. Turn on the gas. Within several minutes, the bird calmly loses consciousness and dies.

Drug Overdoses

Any drug overdose must be administered by a veterinarian. This is a humane method that may suit hobby and pet bird owners best, but it can be expensive for even modest-size flocks. (For more information on euthanasia of turkeys, contact the National Turkey Federation.)

Disposal of Dead Birds

A good health program includes proper disposal of dead birds. There shouldn't be much mortality in small-flock production, but dead birds should still be disposed of properly. Some of the best and most practical methods are burial, incineration, and composting. Never allow fur-bearing animals to eat dead birds, as diseases can harbor in these animals and infect your flock.

Burial is convenient and inexpensive. However, be sure to check with the state veterinarian or Cooperative Extension Service to see if burial is legal. Also, be sure to follow the guidelines that the state or

Cooperative Extension Service might provide. Keeping away wild or feral animals from buried poultry or any other buried livestock is very important. Burial may not be allowed in some states or portions of states where the water table is relatively high.

Incineration is also convenient but is more expensive than burial. You must purchase an incinerator as well as the fuel to operate it. Properly incinerated poultry carcasses do not have an offensive odor when the flame or part of the flame is located between the carcasses and the exhaust pipe. Incineration is a biosecure method of carcass disposal.

Composting is an old and natural phenomenon that has recently been used to dispose of dead poultry. Composting involves more labor than burial or incineration; however, it is a biosecure and environmentally friendly method of disposal. Composting does not involve rotting carcasses, and if done properly, there is no offensive odor.

If the flock is large enough to generate a significant number of dead birds, a compost bin can be constructed. Make this bin small if you plan to empty it with a shovel, or about 6 feet by 6 feet (1.9 × 1.9 m) or larger if you're using a tractor and bucket. More information on composting can be obtained from your local Cooperative Extension Service office.

Rodent Control

A successful rodent-control program consists of three equally important phases: exclusion, habitat modification, and poison baiting.

Exclusion prevents rodents from entering the premises or building. This is difficult but worth the effort in helping to keep down rodent numbers. To exclude rodents, seal all holes and openings with concrete, sheet metal, or heavy-gauge hardware cloth.

POINTS TO REMEMBER ABOUT BAITING

Maintain a constant supply of clean, dry poisoned bait. Check daily and remove any wet or soiled bait. Bait should be maintained in the best possible condition to ensure acceptance by rodents. In many cases, an effective rodent control program can save you money in feed alone.

SINGLE- AND MULTIPLE-DOSE POISONS

Rodenticides are classified into two groups: single dose and multiple dose. Single-dose poisons quickly knock down rodent numbers but may cause bait shyness. Bait shyness occurs when a rodent does not return to the bait after consuming a nonlethal dose, thereby reducing effectiveness. Multiple-dose baits are usually anticoagulants and require repeated ingestion during a 4- to 5-day period to be lethal. Because they are slow acting, multiple-dose poisons do not normally cause bait shyness. Some of the more potent anticoagulants, such as bromadiolone and brodifacoum, cause death after a single feeding.

Habitat modification involves changing the environment of the building or premises to one that is less suited to rodents. The goal is to eliminate sources of food, water, and shelter for rodents.

- Eliminate food sources by cleaning and removing all feed spills, keeping all feed storage areas neat and clean, cleaning feeders and removing spoiled feed, and removing empty feed bags.
- Eliminate water sources by checking all waterers and plumbing for leaks and removing any containers that collect rainwater.
- Eliminate shelter by removing unused equipment, including feeders, waterers, and garbage cans, and removing all vegetation from around buildings; either mow closely or kill weeds or grass.

Poison baiting is the only phase of rodent control that rapidly and consistently reduces rodent populations. However, its effectiveness is reduced when the exclusion and habitat phases of the program are not done efficiently. Commercially available baits contain grains mixed with one of several types of rodenticides.

Bait stations are essential to a successful rodent-control program because they provide a sheltered, secure place for rodents to feed; protect bait from dust, moisture, and weather; allow regular inspection of the bait to make sure rodents are using it; and keep other animals and children out of the bait.

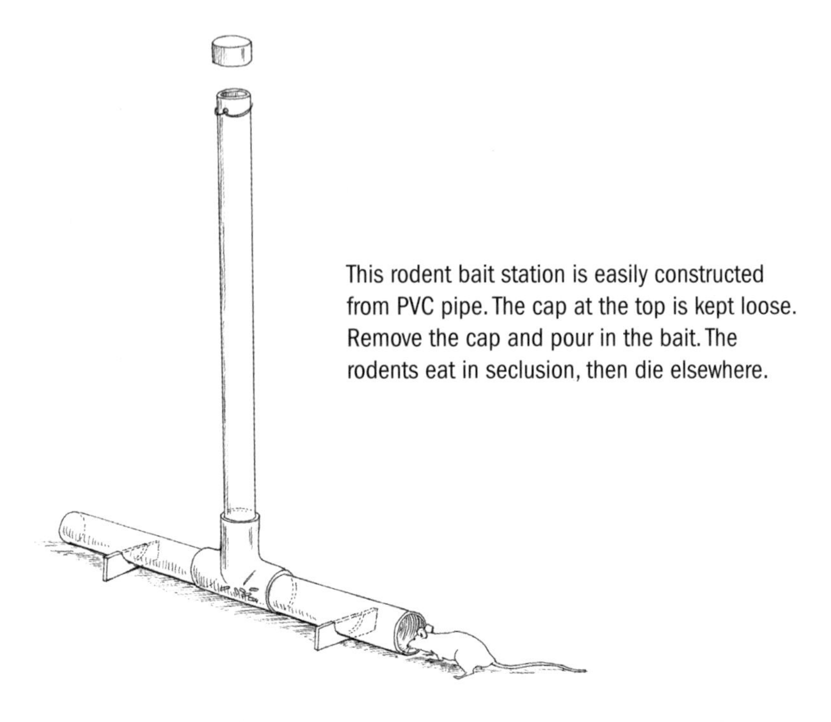

This rodent bait station is easily constructed from PVC pipe. The cap at the top is kept loose. Remove the cap and pour in the bait. The rodents eat in seclusion, then die elsewhere.

Other methods of rodent control include traps, glue boards, fumigants, and tracking powders, but none of these is as practical for controlling rodents as baiting. Baiting tends to be most effective as a preventive measure in the late fall, after harvest and once the first cold snap has arrived. At this time the farm can be rid of rodents and reinfestation is unlikely to occur again until the following fall.

◆ ◆ ◆

Boredom and stress are unhealthy for turkeys. A turkey develops a bad habit like picking because it is looking for something to do. The old guys used to tell me that turkeys will die just from boredom. Furthermore, a bird that carries a burden of several stress factors is highly susceptible to coming down with a disease, if any disease organisms are present.

Master Breeder Frank Reese, Jr., says if a bird is gradually exposed to some of the stress factors, it will successfully adjust to them. The best example of this, of course, is the turkeys' ability to adjust to temperature changes. Growers of the past always knew this.

Advice from Turkey Breeders

ORGANIC AND HUMANE

Upperville, Virginia

SANDY LERNER believes strongly that the best food products come from the best farming practices. To that end, she and her staff at Ayrshire Farm have, for about 8 years, produced approximately 1,200 turkeys each year — raised on pasture, fed organic feed, and humanely handled and slaughtered.

Ayrshire Farm raises Broad Breasted White, Narragansett, Midget White, Bourbon Red, and Bronze turkeys. The poults are purchased each year, but Lerner's commitment extends to slaughtering the turkeys at the farm's own slaughter facility and providing the products for sale to the public at her pub, Hunter's Head Tavern in Upperville, Virginia, and her butcher shop, Home Farm Store in Middleburg, Virginia. Products include whole turkeys for the holidays, smoked or roasted turkey breast, deli meat, pot pie, and a much acclaimed "hot brown sandwich" available at the tavern.

Lerner believes there is value in supporting organic production, heritage breeds, and humane practices that allow the animals to express their natural behaviors. Dave Stevens, a chef and manager at Ayrshire Farm, remarks, "Turkeys love being outside. We could hardly keep them in if we wanted."

THE WEDDING GIFT

Owatonna, Minnesota

WHEN DUANE AND PHYLLIS URCH were married in 1958 they received a pair of Norman Kardosh–bloodline Narragansett turkeys as a wedding gift. This pair formed the foundation of a turkey flock that is still a part of their lives today.

When asked what attracted him to turkeys, Duane responds, "I have just always enjoyed them." This enjoyment has spread: in partnership with his daughter and son-in-law, Julie and Tim Turnland, he operates Urch/Turnland Poultry.

Urch/Turnland Poultry has an impressive collection of rare and hard-to-find chickens, waterfowl, and seven varieties of turkeys, including: Beltsville Small White, Black, Bourbon Red, Bronze, Royal Palm, Slate, and, of course, Narragansett. Duane says the purpose of Urch/Turnland Poultry is to share heritage breeds, and this is done largely through selling hatching eggs.

When asked what advice he would like to share with a beginner, Duane said, "Don't raise the turkeys with chickens — this way you avoid a lot of problems."

FATHER AND SON

Polkton, North Carolina

GARY SIKES, HIS WIFE, KELLY, his eleven-year old son Carson, and Carson's two siblings live in south central North Carolina — less than an hour from the city of Charlotte — and operate Bountiful Harvest Farm NC. The family started looking at food differently after Gary had some health challenges back in 2001. After attending a poultry coaching clinic organized by Jim Adkins, they decided to venture into poultry production — not only for the healthier product that pasture-raised poultry could offer their family, but to make this hard-to-find product available to other health-conscious families as well. Another thing that attracted them to turkeys, they say, was that "they are big, beautiful birds." So in 2011 Gary and Carson purchased some Bourbon Red turkey poults.

Gary focuses on profitability and says that their plan includes selling some turkeys for meat while maintaining a breeding flock and offering quality poults for sale to other farmers. They sell whole turkeys for the Thanksgiving market and poults in the spring through the Internet, at farmers' markets, and by word of mouth.

When asked what advice has helped them most, Carson, who Gary says does most of the work, replies, "How to care for the poults: clean water, good feed, proper housing, and good fences." Gary adds, "From Jim Adkins and Don Schrider, I've learned the importance of quality stock and good breeder selection."

A LIFETIME WITH TURKEYS

Bastrop, Texas

IN 2012, TOM WALKER turned eighty-five years young. At age nine, Tom got started with turkeys when his paternal grandmother, a turkey producer, purchased for him a setting of nine eggs from a "new" variety, the Broad Breasted Bronze. This was in 1936 and broader breasts had just been developed in some lines of Bronze turkeys. Before premixed feeds were available, Tom's grandmother fed young poults a mixture of boiled eggs, green lettuce, and green onion chopped very fine. "That was a pretty good diet then," he says, "and today we know it is still a good diet."

Throughout his young life, Tom's turkey flock provided him with spending money, paid for his first horse, and helped pay his college expenses. He has had turkeys ever since, even during a couple of years spent in Afghanistan beginning in 1969. In 1985, Tom and his wife, Marg, bought their current homestead, of which 1.5 acres is devoted to turkeys.

Tom is well-known for his work saving the rare Regal Red variety of turkey from extinction. He secured his first Regal Red turkeys in 2000, after a 10-year search. The birds needed new blood and Tom, always an experimenter, crossed in some other color varieties of turkey. The result was improved vigor for the Regal Reds, the creation of a new variety, the Harvest Gold, and the resurrection of the previously extinct Black-winged Bronze turkey in a midget size. Tom's work with turkeys is fueled with generosity — the eggs from his flock are a local favorite at the emergency food pantry, and he shares his stock in hopes of encouraging more people to carry on these rare turkeys.

His advice to beginners: "Make sure the poults eat — they need to be taught to eat. If they are with a hen turkey, keep the family in a small area as the poults are very fragile when very young." When asked why he raises turkeys, Tom answers, "FUN!" He says, "They are smart birds and exciting to work with."

AN EXHIBITION ADVENTURE

Englewood, Ohio

SINCE 2010, AT THE AGE OF SIX, ALISHA SPRINGER has shown turkeys with her dad, Brian Schneider, and uncle, Joe Brown. Alisha lives in a small town and it is with great pride that she shares her opinion that turkeys are much better than chickens: "I like how they gobble and how they are smart."

The family raises a few Broad Breasted White and Royal Palm turkeys each year. While they do sell some canned turkey, the real product is the fun of working together with her dad and uncle on a shared family project. She says Denese Newman and Neil Grassbaugh are two of the mentors who have helped with the turkeys.

What advice do Alisha's mentors share? Alisha says, "Make sure you have plenty of room and plenty of money to feed those turkeys." The family raises their turkeys in roomy pens to keep them in top feather condition. She also says, "You can make good money winning Champion Turkey at the fair!" The family shows as B and S Showbirds — next time you are at a fair in Ohio, check to see if Alisha is there with her dad and uncle and a Champion Turkey.

BEGINNER'S LUCK

Hudson, New Hampshire

ONE DAY SUZANNE ROARK visited a friend who had just added turkeys to her farm. She could not believe how beautiful the turkeys were and left feeling quite jealous. So in 2011 Suzanne's friend included five Narragansett turkey poults for Suzanne in her order from Sandhill Preservation Center — thus starting an adventure in turkeys. Suzanne does not have a farm; in fact she has less than an acre and lives only 1 hour from Boston. But the turkeys live comfortably in pens on her property. She keeps only a trio but each year lets the hens raise broods of poults.

Why does she raise turkeys? For meat? To sell breeding stock? "I raise them because they are pretty — really, really pretty," says Suzanne. "I find them clever, good layers, and good mothers." Although her chickens are better at scratching and composting their own manure, she does use the turkey manure in her gardens, "Corn loves fresh manure."

Asked for any tips, Suzanne remarked that most books on turkeys are written with the broad-breasted turkeys in mind and she finds those turkeys difficult to manage. She says the heritage turkeys are totally different. "The poults don't like heat lamps and don't use them much after the first week. The adults like to be outside and prefer to sit atop their house — even in winter."

Her most stunning revelation happened when the turkeys found a way out of their pen. She says, "They worked together looking for a way out. Once one turkeys found a weak spot in the cover, they each took their turn exiting — the group staying together till every turkey was out. It was amazing."

THE GODFATHER OF HERITAGE TURKEYS

Tampa, Kansas

IF YOU HAVE EVER WONDERED HOW a market for pasture-raised, heritage turkeys came about, the answer leads directly back to a fourth-generation turkey farmer from Kansas named Frank Reese, Jr. Frank claims he got started in turkeys by being born — which may not be far from the truth — and always enjoyed and was fascinated by turkeys. He had the privilege of having many great turkey mentors, like Cecil Moore, Hy Patton, Sadie Lloyd, Tommy Reece, Rolla Henry, Agnes Trow, and, most important of all, Norman Kardosh. From these folks Frank learned how to raise, value, and respect these birds.

He also watched as industrial changes to indoor production began to cause a decline in heritage turkeys. He felt he must act to help consumers see the value in these heritage turkeys before they became extinct. Then in 1997 the American Livestock Breeds Conservancy (ALBC) conducted a census that revealed, just as Frank knew, that heritage turkey varieties were close to extinction. Through a partnership between Slow Food USA, ALBC, and Frank, several hundred heritage turkeys were raised for market and supplied to food writers and foodies for tasting. The results came in — heritage turkeys tasted better!

Today on his Good Shepherd Turkey Ranch, Frank and several local farmers raise some 12,000 turkeys per year for market. Frank believes the key to a successful turkey venture lies in the stock. “People do not realize the importance of good bloodlines in producing a quality product with consistency,” he says. “To stay profitable, you must have quality stock and produce a number of different products. Not every turkey makes grade A — you have to have a way to sell those that don’t.” Good Shepherd offers whole turkeys for the holiday market, but they also sell ground turkey, whole breast, sausage, cutlets, live poults, hatching eggs, and some started stock.

When asked for the best advice he received from his mentors, Frank replies, “Stop, watch, pay attention. The turkeys will show you what they need if you simply watch them.” What advice would he offer a beginner? “The last thing you should do is buy the turkeys,” he says. “Figure out the enterprise first — source, feed, processing, and market, and get your equipment.

“Turkeys will die of boredom,” he adds, “but sunlight will fix most things. Get your turkeys outside and you will have fewer problems.”

So why does Frank Reese, Jr., raise turkeys? “Preservation of known old genetic lines. And I enjoy watching turkeys running, jumping, flying — just being turkeys.”

Talking Turkey: A Glossary of Terms

Air cell. Air space in the egg, usually in the large end

Air sacs. The hollows in avian bones which are connected to the nasal system

Albumen. The white of an egg consisting of thick and thin layers

Allantois. Respiratory and excretory organ of bird embryos prior to lung development and activation

Ambient temperature. Actual outside temperature

Amnion. A membranous sac enclosing and protecting the embryo that holds the amniotic fluid

Androgen. A sex hormone produced in the testes and characterized by its ability to stimulate the development of sex characteristics in the male

Anterior. The front part

Antioxidant. Compounds that reduce free radicals in the body; also, compounds used to prevent rancidity of fats or the destruction of fat-soluble vitamins

Avian. Of or pertaining to birds

Bacteria. Microscopic single-celled organisms

Beak. Upper and lower mandibles of chickens, peafowl, pheasants, turkeys, and so on

Beak trimming. Removal of the upper and/or lower tips of the beak to prevent cannibalism and improve feed efficiency

Bits/Rings. Attachments for mandible to prevent cannibalism

Blastoderm. The collective mass of cells of a fertilized ovum from which the embryo develops

Blastodisc. The germinal spot on the ovum from which the blastoderm develops after the ovum is fertilized by the sperm

Blinders/Specks. Attachments for upper mandible to partially block vision. Used to prevent cannibalism

Bread. The tuft of hair attached to the upper chest region

Breast. The forward part of the body between the neck and keel bone

Breast blister. Swollen, discolored area or sore in the area of the keel bone

Brooder. Heat source for starting young birds

Broodiness. Tendency toward the maternal instinct that causes females to set or want to hatch eggs

Bursa fabricious. A glandular organ located dorsally to the cloaca, important to the immunology of the bird, which regresses as the bird matures

Candle. To determine interior condition of the egg through the use of a special light in a dark room

Cannibalism. In the poultry industry, this term refers to the habit of one bird's picking another to the point of injury or death. Can occur as toe picking, feather picking or pulling, vent picking, head picking, or tail picking

Caponization. Surgical removal of the testes from a bird

Carbohydrate. A class or type of nutrient that serves as an energy source and is derived from plant sources, such as grain

Caruncles. The fleshy, nonfeathered area on the neck of a turkey

Ceca. Two blind pouches located at the junction of the lower small intestines and the rectum that aid the digestion of birds, especially when fed highly fibrous diets

Chalazae. Prolongations of the thick inner albumen that are twisted like ropes at both ends of the yolk; they anchor the yolk in the center of the eggshell cavity

Chick. Young bird

Cholecalciferol. Vitamin D3 needed for the absorption and deposition of calcium

Chorion. A membrane enveloping the embryo, external to and enclosing the amnion

Chromosomes. A series of paired bodies in the nucleus of a cell that contain DNA and are responsible for hereditary characteristics; constant in any one kind (species) of plant or animal

Cloaca. In birds, the common chamber or receptacle for the digestive, urinary, and reproductive tracts

Coccidiostat. A drug used to control or prevent coccidiosis, a disease of poultry caused by protozoa

Confinement rearing. Rearing of animals in an enclosed or semienclosed building, such as a barn or shed

Crop. An enlargement of the esophagus in which food is stored and prepared for digestion

Crumble. Form in which some feeds are supplied; it refers to animal feed that has been pelleted and then reground or crumbled into small bits

Cull. A bird not suitable to be kept as a breeder or market bird

Culling. The act of removing unsuitable birds from the flock

Cuticle. External waxy covering or coating of the egg

Darkling beetle. A black beetle that also exists as the lesser mealworm, the larval stage of the darkling beetle, which thrives in poultry bedding material such as pine shavings (litter)

Desnooding. Removal of the fleshy appendage from a turkey's head, usually done at the hatchery or on the day of hatching

Dewlap. A loose flap of skin that hangs between the lower beak and throat

Doral. Of, on, or near the back

Down. Hairlike body feathers covering newly hatched poultry, including turkey poults

Dry-bulb thermometer. Expresses a temperature in number of degrees in Fahrenheit or Celsius

Egg. The female reproductive cell (ovum) surrounded by a protective calcium shell and, if fertilized by the male reproductive cell (sperm) and properly incubated, capable of developing into a new individual of the species

Egg tooth. Temporary extension on the poult's upper beak used to crack the shell at hatching

Embryo. An organism in the early stages of development, as before hatching from the egg

Esophagus. The tubular structure leading from the mouth to the glandular stomach

Evaporation. Changing of moisture (liquid) into a vapor (gas)

resulting in the fusion of the cell nuclei

Fertile. Capable of reproducing

Fertilization. Penetration of the female sex cell (ovum) by the male sex cell (sperm)

Flight feathers. The large primary and secondary feathers of the wings

Flock. A group of turkeys

Follicle. A developing yolk on the ovary

Gang. A group of turkeys; most often used to describe a group of toms

Gene. Element in the chromosome that transmits hereditary characteristics

Gizzard. The muscular digestive organ in birds used for grinding the digesta, located between the proventriculus and intestine

Gonad. Gland that produces reproductive cells; the ovary or testis

Hasp. A hinged fastener for a door that is passed over a staple and secured by a pin, bolt, or padlock

Hatchability of all eggs. Number of poults hatched compared with the number of all eggs set

Hatchability of fertile eggs. Number of poults hatched compared with the number of fertile eggs set

Hatching egg. A fertilized egg intended for incubation

Hen. The female of most fowl, including turkeys

Herd. A group of turkeys

Hock. The joint where the shank (metatarsus) and leg (tibia) meet

Hover. Canopy used for brooder stoves to hold down the heat at bird level

Husbandry. Proper and timely care and management of livestock

Incubate. To maintain favorable conditions for the development and hatching of fertile eggs

Jake. An immature male turkey

Jenny. An immature female turkey

Keel bone. Breastbone or sternum

Lateral. Relating to the sides

Litter. Soft, absorbent material used to cover floors of poultry houses

Mandible. The upper or lower bony portion of the beak

Master Breeder. A talented, experienced breeder able to evaluate and select stock for production, soundness, and breed character, based on close observation of animals.

Meleagris gallopavo. The turkey, both wild and domestic

Molt. To shed old feathers, which are replaced by new ones

Morbidity. Illness

NPIP. National Poultry Improvement Plan, initiated by the USDA in 1935 to reduce transovarian diseases through hatchery sanitation and blood testing

Offal. Waste parts or entrails from butchered or processed birds and animals

Oil sac or uropygial gland. Large oil gland on the back at the base of the tail used by the bird to preen or condition its feathers

Ova. The yolks of eggs

Ovary. The female reproductive gland in which eggs are formed

Oviduct. Long glandular tube where egg formation takes place, leading from the ovary to the cloaca; it is made up of the funnel, magnum, isthmus, uterus, and vagina

Oviposition. The act of laying an egg

Ovulation. The release of the yolk from the ovary

Ovum. The female reproductive cell

Parthenogenesis. Initiation of cell division in an unfertilized egg without contribution from the male. It rarely ends in a hatched poult but it has occurred in turkeys; offspring are predominantly male, with a small number of females

Pendulous crop. Crop that is impacted and enlarged and hangs down in an abnormal manner

Pinioning. Permanent removal of the outer wing joint to prevent fighting

Pipping. Young fowl breaking out of its shell

Plumage. The feathers making up the outer covering of birds

Pores. Thousands of minute openings in the shell of an egg through which gases are exchanged

Posterior. Toward the rear

Poult. A young turkey

Poultry. A term designating those species of birds used by humans for food or fiber that can be reproduced under their care, including chickens, turkeys, ducks, geese, pheasants, and pigeons

Preen gland. Uropygial or oil gland found on the back near the tail that secretes oil for application on the feathers by the bird during preening

Primary feathers. The long, stiff flight feathers at the outer tip of the wing

Protozoa. Microscopic single-celled animals, such as those responsible for coccidiosis, histomoniasis (blackhead), trichomoniasis, and hexamitiasis

Proventriculus. True stomach of the bird, located between the crop and the gizzard

Rafter. A group of turkeys

Range area. An area of pasture or meadowland secured for livestock production

Relative humidity. The percentage of moisture saturation of the air; dependent on air temperature as well as the amount of moisture in the air

Roost. A perch on which birds rest or sleep

Secondaries. The large, stiff wing feathers adjacent to the body, visible when the wing is folded or extended

Semen. Fluid secreted by male reproductive organs; the vehicle for sperm transport

Shank. The scaly portion of the leg below the hock joint and between the thigh and the foot

Shell. The hard protective covering of an egg consisting primarily of calcium carbonate, secreted by the shell gland

Shell gland. That portion of the bird's reproductive tract (oviduct) where the shell and cuticle are deposited around the egg; also incorrectly referred to as the uterus

Shell membranes. The two soft fibrous membrane linings that surround the albumen; secreted in the isthmus, they normally separate at the large end of the egg to form an air cell

Shooting the red. A term to describe the stage when turkey poults lose the down on their head, face, and throat and develop caruncles

Snood. The fleshy appendage on the head of the turkey

Sperm or spermatozoa. The male reproductive cells capable of fertilizing the ova

Spur. The stiff, horny structure on the legs of some birds; found on the inner side of the shank

Strain. Group of birds within a fowl variety of any breed usually with the breeder's name that was reproduced by closed flock breeding for five or more generations

Strut. Mating dance of the male turkey

Testes. The male sex glands (plural) (testis = singular)

Tom. The male turkey

Trachea or windpipe. That part of the respiratory system that conveys air from the larynx to the bronchi and to the lungs and air sacs

Uterus. Organ in female mammals in which the developing embryo is nourished; birds do not have a uterus — see **shell gland**

Vagina. The section of the oviduct that holds the formed egg until it is laid; located between the shell gland and cloaca

Variety. A subdivision of breed usually distinguished by either color or color and pattern, also and/or comb type in chickens

Vent or anus. The external opening of the cloaca

Ventral. Of or relating to the lower part of the body, such as the breast or keel

Vitelline membrane. The membrane that surrounds the yolk

Wet-bulb temperature. Device to measure moisture or water vapor in the air

Yolk. Ovum, the yellow portion of the egg

Definition of a Heritage Turkey from
The American Livestock Breeds Conservancy

All domesticated turkeys descend from wild turkeys indigenous to North and South America. They are the quintessential American poultry. For centuries people have raised turkeys for food and for the joy of having them.

Many different varieties have been developed to fit different purposes. Turkeys were selected for productivity and for specific color patterns to show off the bird's beauty. The American Poultry Association (APA) lists eight varieties of turkeys in its *Standard of Perfection*. Most were accepted into the Standard in the last half of the nineteenth century, with a few more recent additions. They are Black, Bronze, Narragansett, White Holland, Slate, Bourbon Red, Beltsville Small White, and Royal Palm. The American Livestock Breeds Conservancy also recognizes other naturally mating color varieties that have not been accepted into the APA Standard, such as the Jersey Buff, Midget White, and others. All of these varieties are Heritage Turkeys.

Heritage Turkeys are defined by the historic, range-based production system in which they are raised. Turkeys must meet all of the following criteria to qualify as a Heritage Turkey:

1. **Naturally mating.** The Heritage Turkey must be reproduced and genetically maintained through natural mating, with expected fertility rates of 70–80%. This means that turkeys marketed as "heritage"

must be the result of naturally mating pairs of both grandparent and parent stock.

2. **Long productive outdoor lifespan.** The Heritage Turkey must have a long productive lifespan. Breeding hens are commonly productive for 5–7 years and breeding toms for 3–5 years. The Heritage Turkey must also have a genetic ability to withstand the environmental rigors of outdoor production systems.
3. **Slow growth rate.** The Heritage Turkey must have a slow to moderate rate of growth. Today's Heritage Turkeys reach a marketable weight in about 28 weeks, giving the birds time to develop a strong skeletal structure and healthy organs prior to building muscle mass.

This growth rate is identical to that of the commercial varieties of the first half of the twentieth century.

Beginning in the mid-1920s and extending into the 1950s turkeys were selected for larger size and greater breast width, which resulted in the development of the Broad Breasted Bronze. In the 1950s, poultry processors began to seek broad-breasted turkeys with less visible pinfeathers, as the dark pinfeathers, which remained in the dressed bird, were considered unattractive. By the 1960s the Large or Broad Breasted White had been developed, and soon surpassed the Broad Breasted Bronze in the marketplace.

Today's commercial turkey is selected to efficiently produce meat at the lowest possible cost. It is an excellent converter of feed to breast meat, but the result of this improvement is a loss of the bird's ability to successfully mate and produce fertile eggs without intervention. Both the Broad Breasted White and the Broad Breasted Bronze turkey require artificial insemination to produce fertile eggs.

Interestingly, the turkey known as the Broad Breasted Bronze in the early 1930s through the late 1950s is nearly identical to today's Heritage Bronze turkey — both being naturally mating, productive, long-lived, and requiring 26–28 weeks to reach market weight. This early Broad Breasted Bronze is very different from the modern turkey of the same name. The broad-breasted turkey of today has traits that fit modern, genetically controlled, intensively managed, efficiency-driven farming.

While superb at their job, modern Broad Breasted Bronze and Broad Breasted White turkeys are not Heritage Turkeys. Only naturally mating turkeys meeting all of the above criteria are Heritage Turkeys.

◆ ◆ ◆

Prepared by Frank Reese, Jr., *Owner & Breeder*, Good Shepherd Turkey Ranch; Marjorie Bender, *Research & Technical Program Manager*, American Livestock Breeds Conservancy; Dr. Scott Beyer, *Department Chair*, Poultry Science, Kansas State University; Dr. Cal Larson, *Professor Emeritus*, Poultry Science, Virginia Tech; Jeff May, *Regional Manager & Feed Specialist*, Dawes Laboratories; Danny Williamson, *Farmer and Turkey Breeder*, Windmill Farm; Paula Johnson, *Turkey Breeder*, and Steve Pope, *Promotion & Chef*, Good Shepherd Turkey Ranch.

Recommended Reading

American Livestock Breeds Conservancy. *How to Raise Heritage Turkeys on Pasture*. Pittsboro, NC: American Livestock Breeds Conservancy, 2007.

American Poultry Association. *The Standard of Perfection*. Burgettstown, PA: American Poultry Association, 2010.

American Society of Agricultural Engineers. "Manure Production Characteristics," *ASAE Standards*. St. Joseph, MI: American Society of Agricultural Engineers, 2005 (ASAE D384.2 MAR2005).

Billings, Jim. "Secrets of a Turkey Grower." *Bias Magazine*. Springfield, MO: Nov. 21, 1951.

Gilroy, Paul C. "Recollections on Raising Turkeys." Unpublished work; used by permission, 2012.

Hagedoorn, A. L. *Animal Breeding*. London: Crosby Lockwood & Son, 1962.

Lee, Andy, and Patricia Foreman. *Day Range Poultry*. Buena Vista, VA: Good Earth Publications, 2002.

Marsden, Stanley J., and J. Holmes Martin. *Turkey Management*. Danville, IL: The Interstate, 1945.

Reese, Frank R., Jr. "What the 'Old' Experts knew that you should know about stress and turkeys." Unpublished work; used by permission, 2011.

Sanders, Robert. "In the mating game, male wild turkeys benefit even when they don't get the girl." UC Berkeley News, http://berkeley.edu/news/media/releases/2005/03/02_turkeys.shtml, March 02, 2005.

Sponenberg, Phillip, and Carolyn J. Christman. *A Conservation Breeding Handbook*. Pittsboro, NC: American Livestock Breeds Conservancy, 1995.

Turkey World, Mount Morris, IL: Watt Publishing, 1934–1951.

Resources

Organizations

American Pastured Poultry Producers Association
888-662-7772
www.apppa.org
Excellent organization whose members market their own poultry

American Poultry Association
www.amerpoultryassn.com
America's oldest livestock organization, APA licenses expert judges, maintains the breed standards that help ensure productive size and shapes, and governs showing of poultry. APA helps to preserve all breeds of poultry.

The Livestock Conservancy
(formerly The American Livestock Breeds Conservancy)
919-542-5704
http://livestockconservancy.org
A scientific organization devoted to saving and promoting rare breeds

(continued next page)

Recommended Journals and Websites

Exhibition Poultry Magazine
www.exhibitionpoultry.net
Focuses on showing poultry, but excellent content overall

FeatherSite
www.feathersite.com
Has links to breed info and suppliers of all things poultry

First State Veterinary Supply
www.firststatevetsupply.com
In addition to medications and health-related products, offers consulting on health issues with Peter Brown, a.k.a. the Chicken Doctor, through their site www.askthechickendoctor.com. Brown has worked in the poultry industry and is an excellent source of information regarding diseases, general health, reproduction, and even incubation. Well worth the modest consulting fee.

My Pet Chicken, LLC
www.mypetchicken.com
Online resource for poultry and poultry supplies

Poultry Press
765-827-0932
www.poultrypress.com
Oldest poultry publication in the United States and the main one for showing

PoultryWorld.net
info@poultryworld.net
www.poultryworld.net
In their own phrase, a gateway to the global poultry industry. This site often contains useful articles and links — especially on health.

Index

Build Your Barnyard Knowledge with More Books from Storey

The Backyard Homestead Guide to Raising Farm Animals
edited by Gail Damerow
You can raise chickens, goats, sheep, cows, pigs, honey bees, and more in your own backyard! With expert advice on everything from housing to feeding and health care, learn how to keep happy, productive animals on as little as one-tenth of an acre.

Hatching & Brooding Your Own Chicks by Gail Damerow
A definitive guide to hatching healthy baby chickens, ducklings, goslings, turkey poults, and guinea keets, this has information on everything from selecting breeds, having the proper incubator and setup, sanitary conditions, embryo development, and much more.

How to Build Animal Housing by Carol Ekarius
This all-inclusive guide contains illustrated diagrams and in-depth explanations about building shelters that meet animals' individual needs: barns, windbreaks, and shade structures, plus watering systems, feeders, chutes, stanchions, and more.

An Absolute Beginner's Guide to Keeping Backyard Chickens
by Jenna Woginrich
Absolute beginners will delight in this photographic guide chronicling the journey of three chickens from newly hatched to fully grown, highlighting all the must-know details about chicken behavior, feeding, housing, hygiene, and health care.

What's Killing My Chickens? by Gail Damerow
Keep your poultry safe with this lively, logical guide to sleuthing out backyard coop intruders. Learn to piece together clues predators leave behind, understand their hunting habits, and thwart future attacks with sound protective measures.